Smart Dams: Beyond Visual

Parkar

First Printing, 2024

Table of Contents

1 Introduction

1.1 Research background

Dams play an important role in the economy of any country through the provision of water for domestic, industrial and agricultural use, hydropower generation and flood control. However, when these structures fail, the social and economic impact on a country is significant. In this context, a dam is said to fail if its dam wall collapses suddenly and there is uncontrolled flow release of water which was impounded before. Dam failure often results in loss of life and damage to property and environment.

Several dam failures have been reported around the world, for example, the failure of a 30 m embankment dam that claimed over 250 lives near Sheffield in the United Kingdom in 1864 (Brownjohn, 2007), St Francis Lake Dam (Figure 1.1) that claimed 450 lives in the USA in 1928 (Rogers, 2007), Malpasset Dam in France that claimed 423 lives and left over 700 people homeless in 1959 (Londe 1987, Duffaut, 2013).

Figure 1.1. St Francis Lake Dam after failure (Rogers, 2007)

The damage caused by dam failures to the country's economy has resulted in the need for strategies to ensure that dams are safe. Dam safety is referred to as the protection of public and environment from the effects of dam failure, as well as the release of any or all the retained water behind a dam (Canadian Dam Association, 2017). Dam safety ensures that dams are operated and maintained to minimise the risk of failure. To this end, several countries have come up with regulations and guidelines to ensure systematic dam safety management. These guidelines give the dam owners the responsibility of monitoring their dams (Zhang et al., 2016, ICOLD, 2009). The United Kingdom was the first country to have formal legislation known as the 1930 Reservoir Act that gave dam engineers in the UK the responsibility of ensuring the safety of the reservoirs and monitoring their performance (Charles, 2012). As a result, several countries have developed their dam safety guidelines and regulations under the guidance of the International Commission on Large Dams (ICOLD) committee on dam safety (Bowles et al., 2007, Bradlow et al., 2002). As part of these dam safety guidelines and regulations, dam surveillance is a very important aspect of dam safety maintenance.

Othman, (2006) defined dam surveillance as the examination of the condition of the dam and its appurtenant structures and the review of the operation, maintenance and monitoring procedures and results to determine whether a hazardous trend is developing or appears likely to develop. The objective of dam surveillance is to make a precise and timely diagnosis of the behaviour of dams, to prevent undesirable consequences (ICOLD, Bulletin 158). Among the aspects dam surveillance emphasizes are routine visual inspection and monitoring parameters and devices or instrumentation (ICOLD, Bulletin 158). Visual inspection is the most widely used method of identifying structural defects on dams. This procedure involves the examination of all dam components such as abutments, spillways, galleries and so on for cracks and deteriorations. Visual inspections are carried out every after 3 to 5 years by experienced engineers. While visual inspections play a key role in dam safety, the methodology has some drawbacks which include:- (i) being subjective since the judgement depends on the experience of inspectors (ii) some inaccessible elements of a dam are not inspected and defects could go unnoticed (iii) expensive in terms of manual labour in terms of salaries. To overcome the shortcomings of visual inspections, inspired by the developments of sensors and data analysis techniques, dam safety surveillance through dam instrumentation has been used to complement visual inspections. Dam instrumentation is a process that involves the installation of devices on the dam to measure dam structural responses such as displacements and strains and key operational parameters (pore pressure, water level, temperature) thereby enabling a more realistic assessment of the structural performance of a dam, i.e., is the dam structure fit or unfit for the purpose?

The main objective of instrumentation is to ensure and improve dam safety through the provision of information to :- (i) evaluate whether the dam is performing as expected and (ii) warn of changes that could endanger the safety of the dam (Raghavendra, 2013). The advantages of dam Instrumentation are:- (i) provides quantitative data which is sensitive to structural changes (ii) provides long term, consistent data records which allow for trend detections in the dam performance under different conditions and (iii) collection of data can be automated allowing for dam monitoring to be near real-time (Myers and Stateler, 2008). Dam instrumentation forms part of a broader system known as structural health monitoring which is a process of determining the structural health of a structure. In this book, the term dam structural health monitoring will be used. Dam structural health monitoring (DSHM) is defined as a process based on structural static and dynamic responses and environmental variables, measured by sensors installed on a dam body, foundation and reservoir to evaluate structural health of a dam (Cheng and Zheng, 2013). DSHM is divided into static and dynamic response monitoring.

In static response monitoring. dam static properties such as displacements, stresses, strains, cracks, joint opening are measured using instruments embedded in the dam wall in conjunction with environmental variables (water level, air and water temperatures). The information obtained from static response monitoring of a dam is used to decide on the structural health of a dam.

Dynamic response monitoring commonly known as vibration-based structural health monitoring involves monitoring of dynamic properties or modal properties of the dam. These modal properties include natural frequencies, mode shapes and damping ratios. These dynamic properties contain information about mass, stiffness and damping of the dam structure all of which change when there is deterioration. These can also be used as indicators of dam structural health. In dams, dynamic response monitoring has been ongoing since the 1980s with periodic field tests such as forced vibration tests (Severn et al., 1980, Brownjohn, 1986, Loh and Wu, 1996). Forced vibration tests utilise mechanical shakers to cause vibrations on the dam structures. These field tests are expensive in terms of purchase of the equipment and transportation of the equipment to site since dams are always located in remote areas. To overcome these expenses and with the development of sensor and data acquisition technology, environmental sources such as wind and water waves are used to induce vibrations on the dam and responses are measured in form of accelerations. Field tests that depend on environmental sources to induce vibrations on structures and do not require any shakers are known as ambient vibration tests. Therefore, since the 1990s, ambient vibration testing has been a preferable technique to carry out dynamic monitoring of dams. In the earlier years, once-off ambient vibration tests were carried out on dams, for example, Contra Dam in Switzerland (Brownjohn, 1986), Hermitage Dam in British Columbia (Kemp, 1996). In the 2000s, the frequency of ambient vibration testing increased, for example, on Mauvoisin Dam (Darbre et al., 2000), Roode Elsberg and Kouga Dams (Moyo and Oosthuizen, 2010), Contra Dam (Oliviera et al., 2012) to mention but a few. Ambient vibration testing has now become a practice in dam safety management (ICOLD, 2018). This is because responses measured by the accelerometers on a dam site can be used for the extraction of dynamic properties through a process called operational modal analysis (OMA). In OMA, output only accelerations are used to extract the dynamic properties modal parameter estimation techniques such as frequency domain decomposition and stochastic subspace identification techniques etc. (Zhang et al., 2005). Darbre et al., (2000) and Bukenya et al., (2014) indicated that changes in dam natural frequencies were due to change in water level. Although ambient vibration tests are becoming common practice in dam safety management, there is lack of enough knowledge regarding the relationship between the dynamic properties of dams and the environmental variables using long term monitoring data.

In DSHM, much emphasis has been on the static response monitoring with the objective of (i) understanding the static behaviour of dams under different operating and environmental conditions, (ii) calibration and validation of finite element models and (iii) prediction future behaviour (Loh et al., 2011, Bukenya et al., 2014, Salazar, 2015).

Based on current technical literature available dam structural health monitoring systems are either based on static responses or dynamic responses. As a result, there is a need for a study that combines both static and dynamic responses to get a holistic structural performance assessment of dams. Therefore, this research aims at developing a holistic approach in tracking both static and dynamic response of the dam under environmental and operational conditions to understand the structural behaviour of dams.

1.2 Research problem

Safety evaluation of concrete dams is an important practice in dam engineering since existing concrete dams are ageing and deteriorating due to alkali-silica reactions and any past extreme load events such as floods. If these structures are not well looked after, failure can occur leading to loss of lives and property to people living downstream. To ensure the safety of these existing dams, dam structural health monitoring is necessary. DSHM involves the collection of data from sensors installed on the dam wall. Data collected includes dam responses (deformations, stresses, strains, natural frequencies) and environmental loads (water level, air and water temperature). The analysis of the collected data involves a comparison of dam responses with behavioural models so that any deviation from the normal behaviour can be detected. The commonly used behavioural models in the analysis of the collected dam monitoring data include - deterministic, statistical and hybrid models.

Deterministic models utilize numerical methods such as the finite element method in the prediction of dam responses based on physical laws. These models play an important role in the assessment of dam safety during the first filling of the dam since there is not enough data to build data-driven models. However, deterministic models have disadvantages of being computationally expensive and the degree of uncertainty in the characterization of materials and the complexity of the dam structure (Barzaghi et al., 2018). Data-driven models use monitoring data directly to establish the relationship between dam responses and environmental loads through regression analysis. These models have advantages of being simple and fast in execution (Milovojevic et al., 2013). Hybrid models are models where deterministic models have been adjusted to fit the measured data (Salazar et al., 2015).

The shortcomings of deterministic models and the availability of dam monitoring data have made data-driven models a preferable methodology in understanding the behaviour of dams. In dam engineering practice, these models have been used for decades as a complement to visual inspection in making decisions concerning the safety of dams, especially when there is enough monitoring data. Under data-driven models, statistical models are the widely used methods by dam practitioners. In statistical modelling, relationships between environmental loads (water levels, air and water temperature) and dam responses (deformations, strains) are built for purposes of prediction of future behaviour and detection of anomalies. In the past decades, much focus has been on the development of dam deformation-based statistical models such as multiple linear regression (Bukenya et al., 2014, Salazar et al., 2015).

In the last 20 years, focus in the analysis of data collected from dams has shifted to the use of data-driven machine learning algorithms such as artificial neural networks (ANN), support vector machines (SVM) and boosted regression trees (BRT). Some of the case studies where these algorithms have been utilized in the analysis of data include the use of auto-associative neural networks on Fei-Tsui dam (Kao and Loh, 2013), neural networks on Alto Rabagao (Mata, 2011), support vector regression on Djerdap 2 Dam (Rankovic et al., 2014) to mention but a few. More examples of case studies can be found in Bukenya et al., (2014). In Bukenya et al., (2014) and Salazar et al., (2015) it can be seen that much emphasis was put on the prediction of dam deformations and no information is available on the relationship between the dynamic behaviour and environmental loads.

Gaussian process regression (GPR) is a new machine-learning algorithm based on Bayesian theory that has not been used extensively in dam monitoring. GPR offers advantages of being simple to use, flexible and self-adaptive in the choice of hyperparameters and handles nonlinear problems. Therefore, this book intends to develop the relationship between dam natural frequencies and environmental loads. Furthermore, while using GPR models in statistical modelling, kernel functions are key in obtaining good results. Therefore, this study also focuses on the choice of the kernel function in the determination of the relationship between natural frequencies and environmental loads.

1.3 Research question

The research background of this study and research problem has shown that there is a need to continuously monitor dams (loads and responses) to ensure dam safety. Therefore, the research question this book intends to address is:-

What is the effect of environmental loads (water level, air and water temperature) on the dynamic properties?

1.4 Research objectives

Although monitoring of dams has been done for over 20 years, an integrated approach of monitoring both static and dynamic behaviour of dams has not been reported. The major objective of this research is to exploit ambient vibration monitoring and deformation monitoring in structural health monitoring of dams by:-

1) Establishing the relationships between loading (water level and air and water temperature) and the dynamic properties (natural frequencies)
2) Predicting natural frequencies and deformation from monitoring data using data-driven models
3) Using the results obtained in objective (2) for detecting anomalies.

1.5 Conceptual framework

The discussion above has shown that the loads subjected to dams have a big influence on the way in which dams respond. This implies that there are existing relationships between the loads (independent variables) and the dam responses (dependent variables) and these relationships are important in dam safety surveillance for prediction of dam behaviour. Figure 1.2 shows the proposed framework that is used in this study.

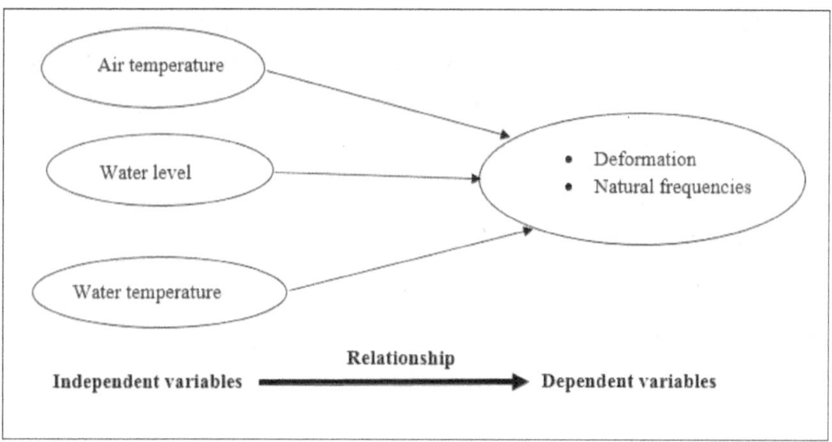

Figure 1.2. Proposed conceptual framework

The extensive review of literature on dam safety and monitoring presented in Chapter 2 suggested that there is a need to develop statistical models that relate the loads on the dams and the dam responses such that the effect of each of the loads can be clearly understood. The above proposed conceptual framework suggests that there is a need to develop relationships between independent and dependent variables shown in Chapter 3. The next section discusses the scope of the study.

1.6 Scope of the study

This work focuses on understanding the behaviour of dams under changing operational and environmental conditions. Roode Elsberg Dam, which is a double curvature arch dam located in South Africa, is used in this research book. To understand this, statistical models based on GPR are developed focusing on which parameters are responsible for the structural behaviour of the dam. After obtaining a baseline model, any deviations from the model will constitute an abnormal behaviour.

1.7 Significance of the study

Dam safety assessment is a very important aspect in any country's economy as it would prevent the timely failure of dams and this will prevent loss of property and lives. Understanding the factors that affect the structural responses which represent dam health can be used in prediction of the future structural behaviour of the dam. This, in turn, can be used to detect any anomalies and decisions can be made in time before a disaster happens.

1.8 Major contribution

The main purpose of dam monitoring is to understand the structural behaviour of concrete dams. This helps in distinguishing between the loads (water level and temperature) on the dam which cause changes and the dam responses (deformations and dynamic properties). Both loads and responses keep on changing over time and these must be monitored over time such that an agreement between the measured responses corresponds with numerical values. Given a large amount of data collected from the monitoring systems installed on the dams, recent advances in the development of statistical models have not been used in dam monitoring. Therefore, there is a need to utilise the recent advancements to analyse data from dam monitoring systems.

The major contributions of this present study are: -

(i) Understand the effect of water level and specifically temperature on the dynamic properties (natural frequencies) of concrete dams.

(ii) Explore the different kernel functions belonging to GPR models in the development of relationships between loads and dam responses. The performance of these models is compared with alternative prediction models such as BRT, ANN, SVR, MARS and MLR.

(iii) Use of robust multivariate statistics (least trimmed squares) in the detection of any outliers in the data collected from the monitoring systems on the dam.

1.9 book outline

This book consists of seven chapters that have been arranged in such a way that it follows the development and methodologies used in this research

Chapter 1 introduces the context of this research by giving a background to the structural health monitoring of dams. Objectives as well as how these objectives will be achieved is presented briefly, research problem, scope, the significance of the study and the main contribution of this research study is also described herein.

Chapter 2 discusses the concept of dam safety including the description of concrete dams, causes of dam failures and citing examples where the failures have occurred. The procedures to be followed in dam safety methodology is presented.

Chapter 3 reviews the concept of structural health monitoring of dams concentrating on the data-driven models that have been adopted in the analysis of dam monitoring data both static and dynamic. Advantages and disadvantages of these methodologies are discussed herein. Case studies where the different analysis methods have been utilised are also presented.

Chapter 4 focuses on the methodology used in the analysis of data collected from the continuous monitoring schemes installed on Roode Elsberg Dam. Gaussian process regression and least trimmed squares models are discussed in detail. Lastly, anomaly detection techniques are presented in the chapter.

Chapter 5 describes the Roode Elsberg Dam and the continuous monitoring systems (deformation and ambient vibration) installed on the dam. Instruments used to measure the environmental variables (water level and air temperature) are described herein.

Chapter 6 presents and discusses the results obtained from the monitoring systems on the dam. Observations of the variations of the dam responses and environmental variables are presented. Also, results from the development of statistical models described in Chapter 4 are discussed.

Chapter 7 summarizes the book, gives conclusions from the research, followed by some recommendations for future research.

2 Concrete dam safety

2.1 Introduction

Water is an important natural resource for all living things. With the increased population and industrialisation, the demand for water has increased over time. This increasing demand for water has led to the need to construct more dams and maintain the safety of the existing ones. Dams are defined as barriers constructed either from earth or concrete across a river to impound water. Dams are constructed to satisfy human needs such as- (i) irrigation purposes (ii) hydropower generation, (iii) flood control and (iv) domestic and industrial use. Therefore, it is necessary for all stakeholders namely: - dam owners and engineers to assess the safety of these important structures to minimise the risk of failure. The objective of this chapter is to provide an overview of dam classification with much emphasis on concrete arch dams and the concept of dam safety surveillance.

This chapter is organised as follows. Section 2.2 discusses how dams are classified, Section 2.3 presents dam failures giving reasons why dams fail and citing some examples. Section 2.4 describes the concept of dam safety and how dam safety monitoring is carried out and issues involved. Lastly, Section 2.5 summaries the chapter.

2.2 Dam classification

Dams can be classified based on their use, hydraulic design, structural design, capacity and materials of construction. Table 2.1 shows the classification of dams based on the above-mentioned categories.

Table 2.1. Dam classification

Classification	**Dam**
Hydraulic design	Overflow, non-overflow
Structural design	Arch, gravity and buttress
Use	Storage, HEP generation, diversion
Construction materials	Concrete, earth fill, rock fill,
Capacity	Small, medium and large

As seen in Table 2.1 above, dams are classified in different ways however in this book, only arch dams will be discussed.

2.3 Arch dams

Arch dams are thin concrete structures whose stability is based on its self-weight and its ability to transmit forces to the valley in which it is constructed. Arch dams have a base thickness of less than 0.6 times its height (Avella, 1993). The classification of arch dams can be according to:-

(i) structural height, i.e. Low arch dams (< 30m), medium arch dams (30-91m) and high dams (>91m).

(ii) Location of the centre and magnitude of the central angle, i.e., constant-centre (variable angle), variable-centre (constant angle) and variable-centre (variable angle).

Figure 2.1 below shows Kouga Arch Dam located in Eastern Cape, South Africa.

Figure 2.1. Kouga Concrete Arch Dam

Concrete arch dams are one of the most important civil engineering structures in any economy. This implies that their safety is important because if they fail, the resulting consequences are fatal. Section 2.4 discusses the concept of dam safety.

2.4 Dam safety

2.4.1 Introduction

Dams provide very important services to the societies they are built-in, however, these services can be accompanied by hazards in case a dam fails. Dams used to be constructed in remote areas where there was no human settlement, however with the growth in population, there is population occupation downstream of these structures. This makes it a concern in case there is potential dam failure. Furthermore, there are so many dams around the world that are ageing and are showing signs of deterioration which can result in dam failure. There have been reported cases around the world (see Section 2.4.2) where dams have collapsed and the resulting consequences have been so costly in terms of loss of lives and property. To avoid these calamities, people who are responsible for these dams have to ensure that dams conform to safety needs. USACE (2011) defines dam safety is an art and science that ensures the integrity and viability of dams such that they do not present unacceptable risks to the public, property and the environment.

A dam is considered safe if it fulfils the following requirements (Medeirros and Torres, 2006):-

(i) support loading conditions over the design life span

(ii) perform its intended purposes to the population.

The objective of dam safety is to protect people, property and the environment from the harmful effects of misoperation or failure of dams and reservoirs (Bowles et al., 2007).

Dam safety management ensures that dams are operated and maintained safely in such a way that dam failures and subsequent consequences are minimised. Pang et al (2006) emphasise that the dam should be evaluated to assess their structural and operational integrity such that a reasonable evaluation of the threatening risk factors is performed.

To ensure dam safety, the following requirements are namely minimization of risk and containing the remaining risk. Mouvet et al. (2001) emphasizes that dam safety goals can be realised through the so-called three tenets (Figure 2.2) namely:-

(i) Structural safety: Making sure that the dam is designed to the latest standards and all the loading and operation conditions

(ii) Monitoring and maintenance: the need for regular inspection of the dam to understand the current behaviour

(iii) Emergency concept: when there is danger, is there a plan in place to get people evacuated and lives saved?

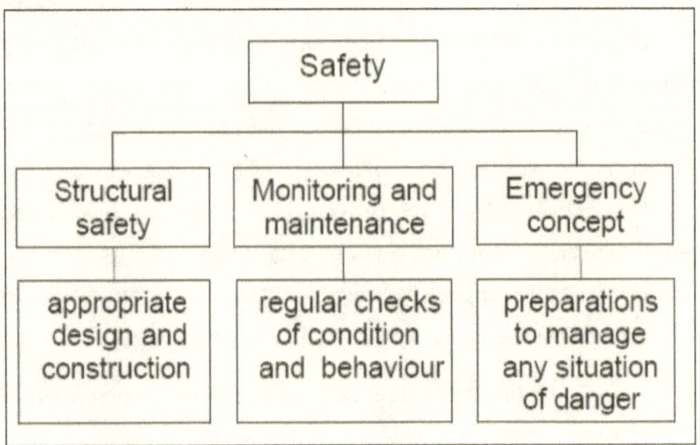

Figure 2.2. Dam safety concept (Mouvet et al., 2001)

2.4.2 Concrete dam failures

"The most appalling catastrophe in the history of mankind, due to failure of the works of man, and one of the worst greatest catastrophes due to any because whatever will make last week forever memorable" (Engineering News. June 8 1889).

Failure of concrete dams can cause damage to human beings and property downstream. In this context, dam failure is defined as a collapse or movement of a part of a dam or its foundation so that the dam cannot retain the stored water (ICOLD, 1995). There have been several reported concrete dam failures around the world and some of the commonest ones are presented here with emphasis on arch dams.

The main causes of failures in concrete dams are: -

- Overtopping of the abutment
- Seepage and piping
- Spillway erosion
- Shear
- Inadequate design
- Foundation

ICOLD reported up to 1965 that around 202 incidents out of 535 were related to total failure of large dam failures. Several concrete dam disasters have been reported causing lots of causalities and loss of property with ICOLD reporting that over 8000 lives have been lost. Examples of concrete dam failures include the failure of Malpasset Dam in France in 1959 that claimed 423 lives and left over 7000 people homeless (Duffaut, 2013, Petaccia et al, 2016), Vaiont Dam which claimed 2600 people in Italy in 1960 (Charles et al., 2011). St. Francis Dam in California which failed in 1928 resulting in the loss of 450 lives and the Austin Dam in Pennsylvania which failed in 1911 with about 80 lives lost (Rich, 2006). Figure 2.3 shows the failure of Gleno Arch Dam in Italy in the year 1923 which had 600 victims. Failure of Bouzey Dam in 1895 (Smith 1994). Table 2.2 gives a summary of some of the concrete dam failures that have occurred around the world, stating the type of dam, when it was built and failed, number of deaths and the causes of failure.

Figure 2.3. Failure of Gleno Dam (Deangeli et al., 2009)

Table 2.2. Examples of some concrete dam failures in the world

Dam	Country	Type	Built	Failed	No. of Deaths	Causes of failure
Walnut Grove	USA	arch		1890	150	overtopping
Gleno Dam	Italy	arch	1923	1923	600	Structural collapse
Malpasset Dam	France	arch	1954	1959	421	Abutment failure
Vaicont Dam	Italy	arch	1960	1963	2600	Reservoir slope slide
Idbar	Yugoslavia	arch	1959	1960		Piping in foundation
Matilija	USA	arch	1949	1965		Deterioration of concrete
Tolla	France	arch	1960			cracking

As seen in Table 2.2 above so many lives have been lost due to dams failing. Therefore, it was important for dam owners and engineers to come up with procedures that will help all stakeholders to ensure that dams are safe. Section 2.4.3 discusses legislations from different countries that were written to ensure that dam failures are reduced and are safe.

2.4.3 Dam safety legislation

Dam safety legislation originated from the United Kingdom after the failure of a Dale Dyke Dam which was a 30m embankment dam leading to a loss of 254 lives near Sheffield in 1864 (Harrison, 1974 and Brownjohn, 2007). The United Kingdom has legislation known as the Reservoir Act 1975 that gives the supervising engineer the responsibility for continual surveillance of a reservoir and dam, including the keeping and interpretation of operational data (Brownjohn, 2007). Several countries around the world have legislation that governs dam safety. For example, Dam Safety Guidelines issued by the Canadian Dam Association (CDA) in Canada, the Dam Safety Act 2000 in India, Dam Safety Act in New South Wales, Dam Safety Act and Decree (319, 2010) in Finland. More information on the different legislation of dam safety in different countries can be found in Bradlow et al (2002).

In South Africa, dam safety legislation is regulated by the following: -

(i) Chapter 12 of the National Water Act 1998 whose purpose is to improve the safety of new and existing dams with safety risk to reduce the potential for harm to the public, damage to property or quality of water resources (van den Berg, 2015).

(ii) Regulations on Dam Safety published in Government Notice R.1560

The purpose of the above-mentioned guidelines is to (Cullis et al., 2007): -

- Outline responsibility of the dam owners regarding the requirements for assessing the safety of a dam with a safety risk.
- List the requirements for both the design report for a new dam as well as the requirements for safety evaluation of an existing dam.

This regulation on Dam Safety in South Africa states that every dam with safety shall be classified according to size (Table 2.3) and hazard potential (Table 2.4). The purpose of this classification is to determine the level of control over dam safety.

Table 2.3. Dam size classification

Size	**Maximum wall height, H (m)**
Small	$5 \leq H \leq 12$
Medium	$12 \leq H \leq 30$
Large	≥ 30

Table 2.4. Classification of a potential hazard

Hazard potential rating	**Potential loss of life**	**Potential economic loss**	**Potential adverse impact on resource quality**
Low	None	Minimal	Low
Significant	$\leq$ 10	Significant	Significant
High	$\geq$ 10	Great	Severe

Table 2.5. Classification of dams using safety risk

Size class	**Hazard potential rating**		
	Low	Significant	High
Small	Category I	Category II	Category II
Medium	Category II	Category II	Category III
Large	Category III	Category II	Category III

Legislation and regulations on dam safety around the world were established and are regularly updated to ensure that dams and their environments are safe. To ensure dam safety, a dam safety program needs to be put in place with activities that will go beyond operational risks as seen in Figure 2.4 (ICOLD, 2009). This program should involve different stakeholders namely: operators, designers and regulators. Each of these stakeholders has well-defined roles and responsibilities in the dam safety program. For example, the operator has the responsibility of operating the dam safety while the designer and regulator are not involved in the day to day operation of the dam.

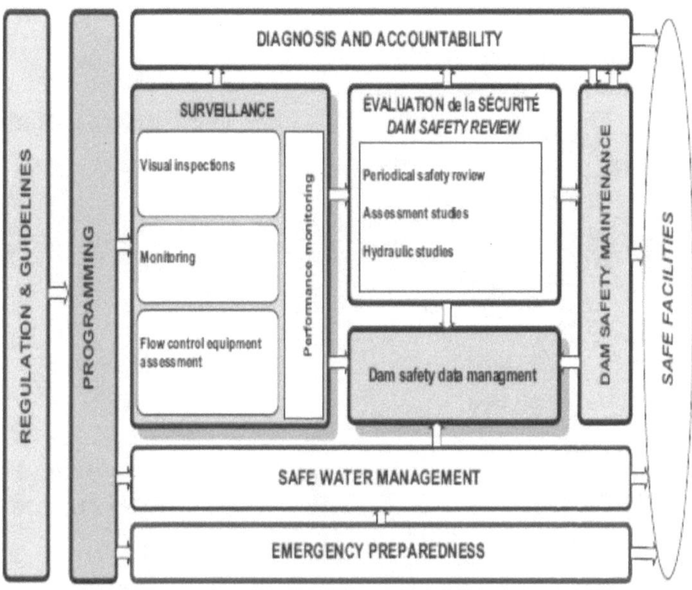

Figure 2.4.Dam safety program (ICOLD, 2009)

The necessity of developing a dam safety program is to ensure that any signs of deterioration on the dam can be captured and if needed some remedial actions are taken. However, in cases where deteriorations or signs of failure are found, there is uncertainty if the dam is entirely safe. To ensure the safety of these dams, there is a need to come up with guidelines that are followed by dam owners. These guidelines are divided into two steps, i.e., risk assessment and dam surveillance (Figure 2.5). Sections 2.5 and 2.6 discuss these two steps.

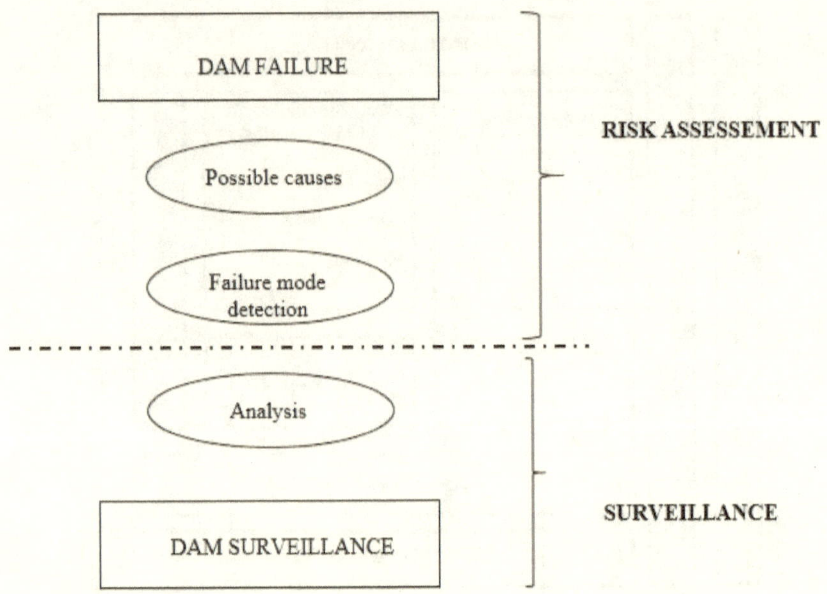

Figure 2.5. Summary of dam safety guidelines

2.5 The risk-based approach in dam safety

The cost of dam rehabilitation proves to be too much for dam owners resulting in the use of analysis of trade-off between cost and risk. Risk analysis plays a very useful role in dam safety programs, i.e., if properly done this methodology provides probabilities of dam failures and consequences which conventional methods may not provide. The threat that a dam causes to human beings and property downstream can be considered in terms of hazard and risk (Singh 1996). According to Singh (1996) hazard is represented by the consequences of failure through possible loss of life and damage.

Risk is defined as a measure of the probability and severity of an adverse effect on life, health, property, or environment (ICOLD, 2005). In simple terms, Zhang et al., (2016) defined risk (R) as a product of probability(P), elements at risk (E), and the vulnerability (V):-

$$R = P \times E \times V \tag{2.1}$$

Risk analysis of dams is part of a dam risk assessment process as shown in Figure 2.6. Figure 2.6 shows that the risk assessment process has a five-step sequence from initiating events to system responses, outcomes, exposure factors, and consequences (Bowles et al., 1998).

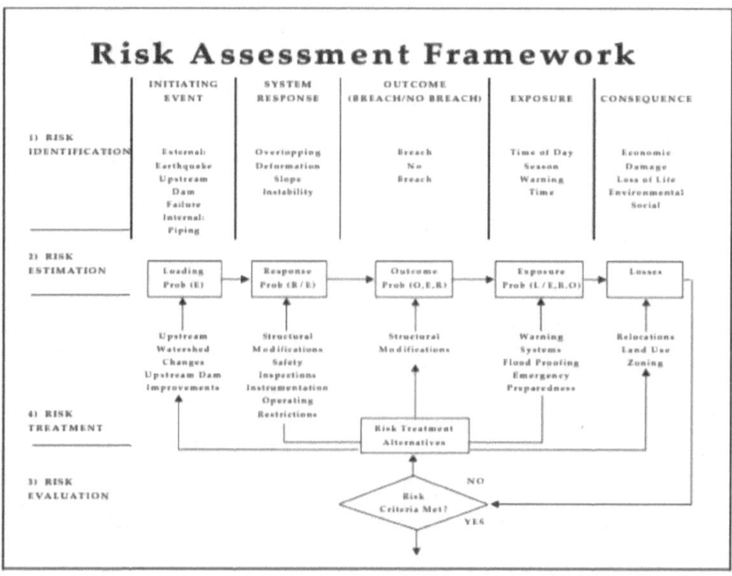

Figure 2.6. Risk assessment framework (Bowles et al., 1998)

Also, Figure 2.6 above indicates four major steps in the process of risk assessment namely: -

- Risk identification is defined as a process of recognizing the plausible failure modes if the dam were subjected to each type of initiating event (Bowles et al., 1998).
- Risk estimation is defined as a process used to produce a measure of the level of health, property, or environmental risk being analysed (Zhang et al., 2016). This process helps in the estimation of loading, system response and outcome probabilities, and the

consequences of various dam failure scenarios and no-failure scenarios, so that incremental consequences can be estimated (Bowles et al., 1998).

- Risk evaluation: The stage at which values and judgement enter the decision process, explicitly or implicitly, by including consideration of the importance of the estimated risks and the associated social, environmental, and economic consequences, to identify a range of alternatives for managing the risks (Zhang et al., 2016).
- Risk treatment: This refers to the consideration of risk reduction alternatives using risk analysis and assessment (Bowles, 1998). Bowles et al., (1998) suggests that risk treatment can be divided into (i) avoid risk which is done before dam construction or process of dam decommissioning, (ii) reduction of the occurrence probability through dam monitoring and surveillance, (iii) mitigating the consequences by the installation of early warning systems on a dam, (iv) transfer of risk which involves the signing of contracts to transfer assets and (v) retaining of the risk.

Risk identification and risk estimations form what is known as risk analysis. Risk analysis uses available information to estimate the risk to individuals or populations, property or the environment, from hazards (Zhang et al., 2016).

2.5.1 Risk analysis

Risk analysis estimates the risk to individuals or populations, property or the environment. Zhang et al., (2016) states that risk analysis provides risk results in assisting risk assessment (whether the risk is acceptable or not and how will the risk mitigation measures work) and risk management (the decision and implementation of risk mitigation measures). Figure 2.7 below shows processes that have to be followed in the process of risk analysis. This process involves the scientific characterization of what is known and what is uncertain about the present and future performance of the dam system under examination (ICOLD, 2005). Risk analysis process estimates both the dam failure probability and possible effects of dam failure.

There are two categories of risk analysis methods: qualitative risk analysis and quantitative risk analysis. Qualitative risk analysis uses word form, descriptive, or numeric rating scales to describe the magnitudes of potential consequences and the likelihood that those consequences will occur.

Quantitative risk analysis is based on numerical values of the probability, vulnerability, and consequences, resulting in a numerical value of the risk (ISSMGE, 2004).

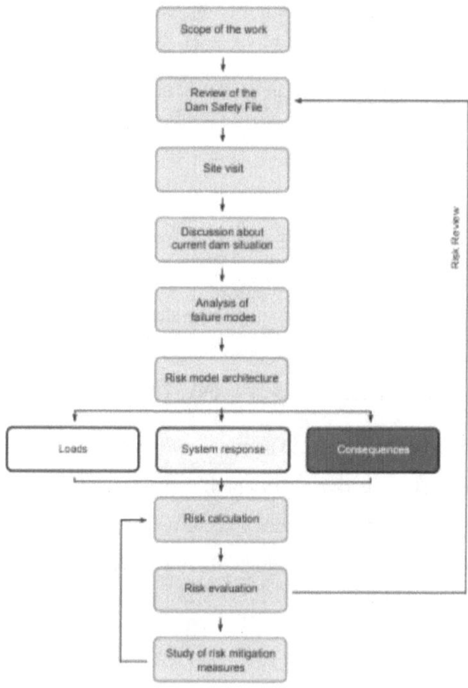

Figure 2.7. Risk analysis process (SPANCOLD, 2013)

The most important step in the risk analysis process is the analysis of failure modes or failure mode identification. A failure mode is defined as a sequence of events that can cause failure or disrupt the function of the dam-reservoir system or part of it (SPANCOLD, 2013). According to SPANCOLD (2013), mode of failure process is associated with a determined loading scenario and has a logical sequence, which starts with a main initial triggering event, is followed by a chain of development or propagation events and culminates in dam failure. The commonly used methods in the identification of failure modes include (Zhang et al., 2016): - failure modes and effects analysis (FMEA), event trees analysis (ETA) and fault tree analysis (FTA).

The identification of failure modes helps in (USSD, 2013):-

- Enhancing dam surveillance programs by focusing on critical parts of the dam.
- Identifying modes of failure that are not easily identified by analytical methods such as slope stability.
- Identifying the most effective dam safety risk reduction measures.

2.6 Dam safety surveillance

Behrouz (2012) defines dam safety surveillance as a two-part process based on period visual inspections of accessible parts of the dam and its surroundings, and systematic monitoring of the dam body and its foundation using instrumentation systems designed specifically for a certain purpose. The objective of dam surveillance is to make a precise and timely diagnosis of the behaviour of dams, to prevent undesirable consequences (ICOLD, 2018).

Dam safety surveillance can be implemented through a proper surveillance program which ensures the safety of ageing dams. The underlying concept of surveillance (Figure 2.8) is: -

(i) The regular inspection to be carried out by qualified engineers

(ii) The need for monitoring a dam throughout its active life by trained personnel

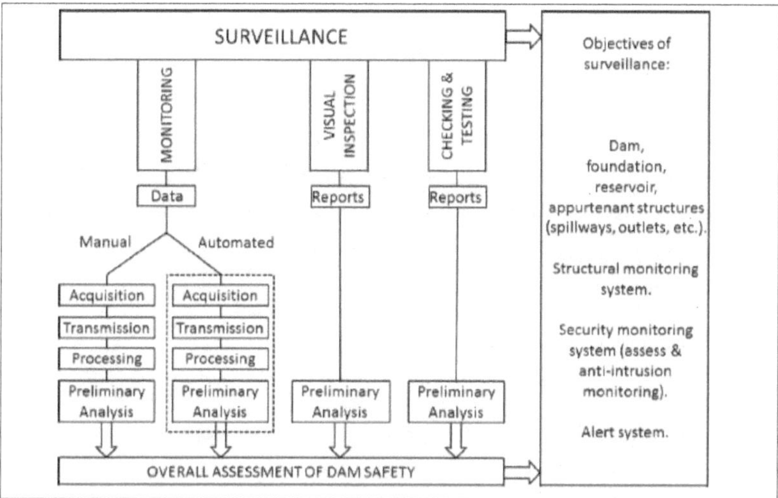

Figure 2.8. The framework shows dam surveillance (ICOLD, 2000)

2.6.1 Visual inspection

Visual inspection involves trained personnel who know dam operations, carrying out regular and careful inspections on the dam such that any damages such as cracks or deteriorations due to alkali-aggregate reaction (Hattingh et al., 2017, Batista and Gomes, 2016) can be detected at an early stage. These inspections are important for dam owners as observations such as cracks, vegetation, leakage and signs of seepage is information which can be used to reconstitute the history of the dam. There are three types of visual inspections namely (Nordström et al., 2019): -

- Routine inspections which occur weekly:
- Inspections which occur biannually
- The in-depth inspection is carried out every 9 years

The above-mentioned inspections detect any deviations from the expected dam behaviour such that an initial judgement can be made on dam safety.

Visual inspections have advantages of being cheap and straightforward for the dam owner to ensure the safety of the structure. However, this methodology has several disadvantages such as high demand for manpower, often inadequate frequency and inaccessibility of critical parts of the structure under inspection. For example, should damage occur on the upstream side, underwater level or contact between dam and foundation rock, they are hard to observe them directly. The consequential shortage of knowledge about the structural condition can result in an erroneous assessment of the condition of the structure and consequently uninformed decisions regarding the maintenance of the structure. To overcome the shortcomings of visual inspections, instrumentation monitoring of dams is used. Section 2.6.2 discusses what is involved in the instrumentation monitoring of dams.

2.6.2 Instrumentation monitoring of dams

Dams will not fail without showing signs of danger. In some of the reported dam failures, ample signs of danger were seen but ignored. In other instances, if monitoring systems had been installed on the dams and results analysed properly, the deterioration could have been detected. Also, the operation of dams involves potential risks of high damages which requires dam safety to be permanently assured by reliable monitoring systems (Cunha et al., 2017).

As part of dam safety, factors such as structural, hydraulic, geotechnical, environmental and operational should be considered during the life span of a dam. Therefore, there is a need for a dam monitoring suite that monitors the geotechnical and structural behaviour of the dam such that dam behaviour and integrity is assessed. Monitoring is defined as the observation of measuring devices that provide data from which can be deduced the performance and behavioural trends of a dam and apartment structures, and the recording of such data (ANCOLD, 2003).

The purposes of instrumentation include (Schurer et al., 2002, Novak et al., 1996): -

(i) Detecting any structural problem on the dam that cannot be observed visually, for example, changes in pore water pressure that can lead to seepage flow problems.

(ii) Analysis and definition of the problem: For example, if the dam is deforming uniformly or is the deformation increasing in some areas relative to others.

(iii) Proving that the dam is behaving as expected: This is important for dam owners and engineers to feel at ease that the dam is behaving as designed.

(iv) Evaluating the performance of any remedial actions: When remedial actions have been carried out on the dam, it is important to see how these perform in ensuring that the dam is safe.

(v) Calibration of numerical models

Monitoring allows a continuous updating of the knowledge about the dam's condition throughout its lifetime and therefore provides a great improvement in the timely detection and correction of eventual deterioration scenarios (Pedro, 1999). This is done through the measurement of loads that dams are being subjected to and the structural responses from the dam because of these loads (Figure 2.9).

The frequency of monitoring is influenced by the following factors (QDMG, 2002):

(i) The consequence of dam failure

(ii) The nature of the behaviour being monitored

(iii) The age of the dam

(iv) The existence of events such as floods, earthquakes which may require continuous monitoring.

(v) Existence of defects identified through dam safety audits

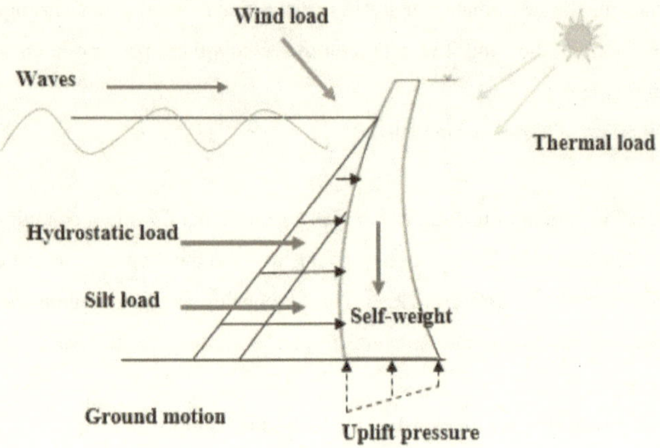

Figure 2.9. Loads acting on the arch dam

The loads and responses of the dam are measured by several instruments as shown in Table 2.6 below. Instruments used in the monitoring of loads and responses should have the following characteristics (ICOLD, 2018): -

- Robust
- Durable
- Simple to use and easy to read
- Insensitive to temperature, humidity
- Precise

Table 2.6. List of some of the commonly measured parameters and corresponding instruments

Parameter	**Instrument**
Water level	Staff gauge
Air temperature	Thermometer
Water temperature	Thermocouple
Concrete temperature	Thermocouple inserted in the dam wall
Deformations	Plumblines, GPS
Strain	Strain gauges
Pore pressure	Pressure gauge
Accelerations	accelerometers

2.7 Chapter summary

Chapter 2 has discussed what concrete dams are and how they are classified with emphasis on rigidity classification. Dam failures including causes, modes of failures and citing examples where dams have failed have been described herein. Dam failures mentioned in this chapter can be avoided if proper dam safety guidelines are followed. Dam safety ensures that dams do not present any risks to the people, property and the environment in which they are built. To mitigate the risks dams may pose due to their failure, all countries around the world have developed dam safety programs. The concept and what a dam safety program entails has been discussed in this chapter. Dam safety programs are usually designed to inspect and monitor the structural integrity of dams. This involves visual inspections and instrumental monitoring. Despite the growing interest in ensuring dam safety through the dam safety programs, there exist some issues that have been identified in this review and need to be addressed. Below are some of the issues that have been identified.

Most of the time, dam engineers rely on reports prepared by dam operators every quarter but do not know how accurate they are. Therefore, engineers need to frequently visit dam sites.

The preparation of dam safety reports is still very labour intensive which may have errors. Therefore there is a need to have intelligent information systems that can help in the reduction of labour effort needed and will improve the quality of information. This will further help engineers dealing with dam safety to evaluate data continuously and short and long term dam safety condition assessment will be made easier.

In conclusion, instrumental monitoring produces a huge amount of data which is normally presented on graphs. However, there is a need to analyse this data in-depth to get as much information as possible. Chapter 3 discusses the methodologies that are used to analyse data collected from dams.

3 Dam structural health monitoring

3.1 Introduction

Dam structural health monitoring involves measurement of loads on the dam (temperatures, reservoir level) and dam responses (displacements, stresses, opening and closing of joints, crack opening, strains). Data collected from dams has to be interpreted such that the load effects on the structural behaviour of dams can be considered separately. This process enables the answering of the following questions (Lombardi, 2004):-

(i) Is the dam safe enough at this time?

(ii) Will the dam be safe at a given time in future?

The analysis and the interpretation of monitoring data is an important activity of a dam safety program. The purpose of this activity is to provide the necessary background about dam behaviour for a better definition of the requirements (data selection, type of models, etc.), to enhance the conceptual understanding and to represent the dam's behaviour.

Structural responses of dams such as deformations and natural frequencies are dependent on hydrostatic pressure measured from dams which are normally affected by factors especially hydrostatic pressure and temperature, mechanical properties of concrete, and alkali-silica reaction (Rankovic et al., 2014). The relationship between the loads on the dam and the structural responses are established through: -

(i) Physics-based models: These models are also known as deterministic models that use numerical methods such as finite element models to predict dam responses utilising physical governing laws. Physics-based models have an advantage in that they can be used during the first filling of the dam which is perceived as the most critical part of the dam's life (Prakash et al., 2018)

(ii) Data-driven models also known as statistical models: formulate the correlation between measured structural responses and the loads. Statistical models define relationships between the current and the past behaviour of the dam. The accuracy of this model depends on how large and how accurate the available data is. This approach offers a solution to the question: Does current dam behaviour correspond to the behaviour observed in the past?

(iii) Hybrid models are physics-based models whose parameters have been modified to fit measured data. These models allow the calibration of FE-models based on the results from the data collected from the monitoring systems. Karimi et al., (2010) proposed a hybrid model based on ANN and finite element modelling of a concrete gravity dam with an empty reservoir for system identification. Data from forced vibration tests was used in developing the ANN model.

(iv) Mixed models: These are developed in such a way that physics-based models are used to predict dam response due to water level effects and a statistical model to predict dam response due to temperature effects.

Dam responses such as natural frequencies cannot be predicted accurately through physics-based models due to limited knowledge on loads, properties of dam foundation materials and physical laws governing the stress-strain relationships. Also, assumptions are made on the geometry and boundary conditions while formulating deterministic models. On the other hand, data-driven models have advantages of correlating any loads and the dam responses. Therefore, this chapter discusses data-driven models that have been used by several authors in the prediction of dam responses, citing the advantages and disadvantages of the models and the case studies where they have been used. This will help in understanding the state of structural health monitoring in dams and guidance to research needs.

3.2 Review of data-driven analysis techniques for static properties

3.2.1 Introduction

The most commonly monitored parameters in dam surveillance are air temperature, reservoir level and deformations. Thus, data analysis seeks to establish the relationship between the measured parameters such that conclusions can be made about the dam behaviour. Several data-driven analysis techniques are being used by engineers and researchers to analyse static properties measured from dams. The analysis techniques are divided into regression-based techniques, time-series based techniques, machine learning-based techniques and other methods.

3.2.2 Regression-based techniques

Regression-based techniques are part of predictive modelling that formulate relationships between dependent variables (for example deformations) and independent variables (water level, air and water temperatures). This section discusses regression-based models that have been commonly used in dam monitoring.

3.2.2.1 Hydrostatic-season time (HST) model

The hydrostatic season time model is a regression-based model that is still widely used by several engineers to assess the different type of observations on dams (Chouinard and Roy, 2006). The hydrostatic season model was developed by Electricity de France (Garabedian et al., 2006). This model interprets observations from the dam and can also identify any significant deviations from normal behaviour. The HST model identifies the relationship between a single dependent variable and several independent variables also known as predictors. The established model allows the prediction of future values of the dependent variable when only the predictors are known. In dam monitoring, the HST model takes into account the hydrostatic load due to water level, thermal load due to water and air temperature and the time effect (Bonelli and Poyet, 2001). These models rely on some basic assumptions: (i) the analysed effects correspond to a period in which the configuration of the structure remains the same. (ii) The response of the dam can be separated in two parts, a) reversible effects due to the variation of hydrostatic level and air temperature, and b) irreversible effects which are a function of time and can be induced by creep, alkali aggregation reaction, or other damage. A general statistical model for the response of an instrument can be formulated as shown in Equation 3.1 (Bonelli and Poyet, 2001): -

$$y = f(h,s,t) = f_h + f_s + f_t + a + \varepsilon \qquad (3.1)$$

Where f_h is the deformation due to the elastic effect due to water level f_s, is the deformation due to the effect of temperature depending on the thermal conditions, and f_t is the deformation due to the effect of time, a is a constant representing the deformation at the beginning of the analysis and ε is the error.

The deformation due to the effect of the hydrostatic load is represented by polynomials depending on the factor h which depends on the height of the water in the reservoir as shown in Equation 3.2 (Bonelli and Poyet, 2001): -

$$f_h = \beta_1 h + \beta_2 h^2 + \beta_3 h^3 + \beta_4 h^4 \qquad (3.2)$$

where $h = \frac{H_{max} - H_i}{H_{max} - H_{min}}$ H_{max} and H_{min} are the historical maximum and minimum water levels respectively while H_i is the measured water level at day i (Bonelli and Poyet, 2001):

The thermal effect is represented by the sum of sinusoidal functions within 1 year (Perner and Obernhuber, 2010). Therefore, the effect of temperature variations can be represented as a linear combination of sinusoidal functions that only depends on the day of the year as shown in Equation 3.3 below (Bonelli and Poyet, 2001):-

$$f_s = \beta_5 \cos(s) + \beta_6 \sin(s) + \beta_7 \sin^2(s) + \beta_8 \cos(s)\sin(s) \tag{3.3}$$

where $s = \frac{2\pi d}{365}$, d is the number of days between the beginning of the year i.e. January 1 until the date of observation (0≤d≤365).

Time effects can be represented in Equation 3.4, where t is the number of days since the beginning of the analysis.

$$f_t = \beta_9 e^{-t} + \beta_{10} t \tag{3.4}$$

βs are the regression coefficients.

The unknown parameters in the HST model can now be solved using multiple linear regression using the least-squares method, i.e. the model is fit in such a way that the sum of squares of the difference of observed and predicted values (Montgomery et al., 2012).

The HST model has been used to analyse data from dams because of the following advantages: -

(i) Frequently provides useful estimations of concrete dam displacements (Tatin et al., 2015).

(ii) The methodology is simple and easily interpretable and used worldwide by engineers (Tatin et al., 2015).

(iii) There is no need for air temperature as periodic functions are used for the thermal effect.

Although HST is one of the most preferred methods of analysing data by engineers, this model has limitations. These limitations include: -

(i) The assumption that hydrostatic load and thermal effect are independent is not true

(ii) There is a loss of prediction accuracy since the actual temperature is not taken into consideration

(iii) Non-linear effects such as closing and opening or cracks are not considered (Zhao, 2003)

(iv) It shows reversible effects in model performance as anomalies.

3.2.2.2 Hydrostatic thermal time (HTT) model

Hydrostatic temperature-time models consider the actual measured temperatures on the dam instead of representing the thermal effect as sinusoidal functions like in HST models. Several approaches have been proposed to improve the modelling of the thermal influence by accounting for real temperature evolution. Statistical models that considered explicitly data from thermometer embedded in the concrete mass have been proposed (Leger and Leclerc, 2007). HTT model resembles the HST model but only the actual temperatures on the dam wall replace the sinusoidal function shown in Equation 3.3. Equation 3.5 shows the temperature variation (Yu et al., 2018).

$$f_s = \sum_{i=1}^{N} b_i T_i \tag{3.5}$$

Where T_i represents measured temperatures from the thermometers installed in the dam wall at different heights and b_i are the regression coefficients to be determined.

The advantage of using concrete temperature measurements directly is that uncertainties, due to heat transfer processes, especially at the boundaries can be avoided. However, HTT has some shortcomings such as: -

(i) the spatial location of temperature sensors which makes it difficult to select thermometers to use.

(ii) Need for thermometers in the dam body that present a phase offset similar to the structural response analysis under analysis (Rico et al., 2019).

(iii) many dams do not have embedded temperature sensors.

These shortcoming has been overcome by using principal component analysis (Mata, 2011 and Mata, 2013) on the entire set of temperatures and dam responses where the principal components with relevance are chosen. The other option is to apply PCA to only temperatures and use principal components in the regression model (Mata, 2013).

3.2.2.3 Thermal Hydro-season time model

Tahin et al (2015) proposed a hybrid model known as Thermal HST (HSTT) which keeps the seasonal function of HST but adds a corrective term which accounts for delayed deviation of the daily air temperature to its seasonal average as shown in Equation 3.6. This method enables a significant reduction of the residual dispersion of the HST model and reduces the anomalies introduced by extremely high-temperature conditions. (Tahin et al., 2015).

$$y = f_h + f_s + f_t + f(\Delta\theta_R) + a + \varepsilon \tag{3.6}$$

Where; $\Delta\theta_R$ represents the thermal displacement induced by the air temperature deviation $\Delta\theta$ This term also enables a better explanation of the displacements measured during abnormal climatic events at the cost of the estimation of air temperature to feed the model.

3.2.2.4 Hydrostatic Season Temperature-Gradient (HST-Grad) model

Tahin et al (2015) developed a hybrid model that involves both mean temperature and temperature gradients across the dam wall thickness also considering water temperatures. This hybrid model is called the HST-Grad model. Tahin et al (2015) estimated the temperature gradients from the air temperature in the downstream face and a weighted average of the air and water temperatures in the upstream one. The HST-Grad model is shown in Equation 3.7 below: -

$$y = f_h + f_s + f_t + f(S, \Delta T_m \Delta T_G) + a + \varepsilon \tag{3.7}$$

where T_M and T_G are respectively the mean temperatures and the temperature gradient across the thickness of the dam. The advantage of HST-Grad is that it improves the prediction of dam displacements due to temperature.

3.2.2.5 Hydrostatic Season State (HSS) model

In the HST model, the time effect cannot be measured directly but it can be regarded as a structural state of a concrete dam that, to some extent, reflects the operation characteristics of the dynamic system of the concrete dam (Li et al., 2014). Li et al. (2014) suggested an improvement of HST model introducing a state-space model in Equation 3.1 calling it the hydrostatic season state model as shown in Equation 3.8 below:

$$y = f_h + f_s + f_t + M_t f_{\theta,t} + a + \varepsilon \tag{3.8}$$

$$f_{\theta,t+1} = \Phi_t f_{\theta,t} + w_t \tag{3.9}$$

Where M_t is known as an observation matrix, Φ_t is the transfer matrix, $f_{\theta,t}$ is a vector that represents the unobservable structural state of concrete at that particular time and w_t is the vector of random disturbance. The relationship between an observable vector and the internal state of the dam is represented by Equation 3.8 while the evolution of the state of a dynamic system of a concrete dam from the previous moment to the present moment plus the transition rule of state variables that become difficult to observe directly in the safety monitoring of the dam during its lifetime is described in Equation 3.9 (Li et al. (2014).

3.2.2.6 Considering delaying effects

Several researchers have shown that dams may respond to certain loads with some delay, for example, the influence of air temperature in the thermal field on the concrete body (Mata, 2011). Salzar (2015) gave a list of studies where moving averages, gradients of the independent variables were used to predict dam responses. However, these studies were not satisfactory until Bonelli and Poyet (2001) proposed a methodology based on the convolution integral of the impulse response function (IRF) and the loadings:

$$y = \alpha \frac{1}{t_0} \int_0^t e^{-(\frac{t-t'}{t_0})} h(t')\partial t' \tag{3.10}$$

where α is a damping coefficient, t_0 is the characteristic time, which depends on the phenomenon, and $h(t')\partial t'$ is the reservoir level at a time t_0. Although the analytical integration of this function is cumbersome, it can be solved using numerical approximation. The advantage of this approach is that the coefficients have physical meaning: the characteristic time provides insight into the lag with which the dam reacts to a variation in the input variable, whereas the damping reflects the relation between the amplitude of the reservoir level variation and that of the pore pressure in the location considered within the dam body.

A similar approach was followed by the same author in the frame of the above mentioned 6th ICOLD Benchmark Workshop Bonelli and Royet (2001). In this case, it was intended to account for the delayed response of the dam in terms of the temperature field, with the final aim of predicting radial displacements.

There are several case studies in literature where HST and similar models (to overcome the shortcoming of HST and also include certain parameters) have been used by dam engineers to analyse data from monitoring systems installed on dams. Below are some of those case studies.

Behrouz (2002) analysed displacement data collected for 24 years from Daniel Johnson Dam in Canada. The hydrostatic season-time (HST) model was used for modelling and forecasting the future displacements of the dam. Two HST models, i.e. Model 1 which approximated temperature using sinusoidal functions and Model 2 that used concrete temperatures were used in the prediction of dam displacements. Behrouz (2002) concluded that the HST Model 2 performed better than the HST Model 1.

Chouinard et al. (2006) discussed the advantages of multivariate statistical analysis to study the long term behaviour of Daniel Johnson Dam. The main objective of this study was to estimate the components of the behaviour of the dam and determine the importance, extent and type of irreversible displacements for 9 years. The hydrostatic season time (HST) model was used to model the long term behaviour of the dam. Investigations of dependencies between displacements over the entire structure were carried out using principal component analysis. Results indicated that the thermal effects, response to reservoir water level and the irreversible effects were the three main components of displacements.

Garabedian et al., (2006) compared the performance of three multiple linear regression models namely HST, HSTT and HTT using data provided by the British Columbia hydro from one of its concrete gravity dams. Water level, temperature and deformation ranging from 1988-1999 were used in the analysis. Results from the analysis showed that HSTT performed best amongst the three models with the highest R^2 value of 0.904.

Huaizhi et al (2008) introduced the so-called resistance (dam structural behaviour and material factors) and source factors (water pressure and temperature) in the traditional HST model development. The proposed model was known as the game model which incorporated the maximum height and elastic modulus of the dam in the HST. Eight years of monitoring data from a dam in China were used in the development of the model. Authors concluded that the proposed dam model performed better than the traditional HST model with an R^2 value of 0.97.

Hong et al (2010) used principal component analysis in the reduction of monitoring data of Chencun arch-gravity dam and then the reduced data was used in the HST model to predict crack opening of using a 10-year data. Water level and air temperature were used as the input variables. Results showed that principal component analysis helps in noise reduction and filtering hence the level of false alarms can be minimised.

Lopes et al (2012) developed dam behaviour prediction models using multiple linear regression and control charts from measured displacements of Tucurui Dam located in Brazil. This study was purposed for proposed measures of how to improve measuring procedures on concrete dams and determining changes in structural behaviour. In an attempt to calculate the difference between the predicted and measured values, control charts were used. Results indicated that the change in the dam behaviour was in the second construction phase of the dam.

Barzaghi et al (2012) described a monitoring scheme of two gravity dams namely Genna Is Abis and Eleonora d'Arborea located in Italy. Crest deformations of the two dams were measured using pendulums and compared with that from the Global Navigation Satellite System. Statistical models based on the HST model were developed on the 20 years and 6-year data for Genna Is Abis and Eleonora d'Arborea Dams respectively. The analysis showed that the maximum amplitude of displacements was in the range of 3-4mm.

Li et al., (2013) proposed an error correlation-based HST model in the analysis of deformation data collected from the Wanfu Dam located in China. The analysis procedure included the use of Augmented Dickey-Fuller tests to test the stationarity of the time series based on Cointegration theory. The error correlation model was compared with the HST model and results showed that the model performed better in fitting and forecasting.

Gu et al (2013) developed a semi-parametric statistical model for crack monitoring of a concrete gravity-arch dam in China. This was the detection of a crack of almost 5m deep and a length of 300m running across the dam wall. Reservoir water level and concrete temperatures were used as predictors to model crack behaviour. The developed model was compared to the parametric model in capturing the aperture of the crack and it was seen that the semi-parametric model performed better with an R^2 value of 0.991 compared to 0.926 of the latter.

Mata et al., (2014) compared two statistical models namely hydrostatic thermal time (HTT) and hydrostatic season time (HST) models in the monitoring and interpretation of radial displacements of Alto Lindoso Dam located in Portugal. Water level, air temperature and concrete temperature and radial displacements for five years were used in the development of the models. Results showed that HTT models performed better than HST as they had a lower residual standard deviation and they do not depend on the cyclic evolution of cyclic thermal load.

Li et al (2014) presented a hydrostatic seasonal state (HSS) model using the expectation maximisation algorithm to extract the time effect deformation from the monitoring data taken from Huaguantan Dam. These time effect deformations were estimated because they represent any changes in material properties, creep and shrinkage effects. Concrete temperatures, water levels and deformations collected for eight years were used in the modelling. The HSS model was compared with the HST model and results showed that HSS models performed better than the traditional HST model in such a way that:- (i) it can effectively estimate the time-effect deformation in the monitoring data analysis of concrete dam deformation, avoiding the possible uncertainties of choosing time-effect expressions; (ii) it has a better fitting and prediction precision than the HST model and (iii) the extracted time-effect deformation generally reflects long-term accumulation, indicating that time-effect deformation increases over time.

Nikolovski and Slaveski (2014) presented results from a monitoring scheme of Sveta Petka Dam located in Macedonia. Sveta Petka Dam was monitored from construction, first filling and during the exploitation period. Results obtained from the analysis of dam data were compared with those from the mathematical model. Results indicated that to avoid any danger especially during the first filling period (critical stage) of the dam, it is important to monitor the behaviour of the dam and that the stresses caused by temperature changes can be larger than those from the reservoir loading.

Tatin et al (2015) extended the HST model to account for both the mean and gradient of the temperature of the dam body. The HST-Grad model takes into account both the mean and the gradient of the temperature in the dam body, considered as a one-dimensional domain. They are estimated from the air temperature in the downstream face, and a weighted average of the air and water temperatures in the upstream one. A similar and more detailed approach was applied by the same authors, called the SLICE model. This model considers different thermal conditions for the portion of the dam body located below the pool level to that situated above, which is not affected by the water temperature. The performance of HST-Grad model was compared with the HST model using displacement data ranging from 2005-2013 collected from 8 French dams namely Eguzon, Izourt, Roseland, Tignes, Vouglans, Bissorte, Gittaz and Sarrans. Results showed that with the HST-Grad model there was a reduction of residual dispersion.

Lazzarotto et al (2016) proposed multivariate statistical models based on control charts and principal component analysis for data collected from Itaipu Dam. The main purpose of the study was to separate the effect of environmental variables on the dam responses and false alarm detection. Analysis of results showed that false alarms on the dam can be reduced using T^2 and Q-statistics.

Song et al (2016) predicted displacements of Xiaowan Dam located in China using the HST model. Data used in the development of the prediction models of the dam was 3 years. The main objective of the study was to develop a damage detection algorithm based on the monitored data. The relationship between the observed data and damage information was formulated using the inversion method. Damage parameters were calculated using the damage information.

Bak (2016) used the HST model in forecasting displacements of Solina Dam located Poland. Feeler gauges installed on the pillars, retaining walls and four galleries were used for displacement measurements. Four different forecasting periods namely: - 10, 12,15 and 18 years were used in the analysis and it was concluded that reliable results can be obtained when monitoring data spans for 10 years measured at a frequency of one reading per month.

Alcay et al (2016) compared displacement data measured by the direct pendulum and geodetic data on Ermenek Dam located in Turkey. The main objective was to develop relationships between the loads (water load and temperature) on the dam and dam response (displacements) and ascertain which data gave best results. Observations indicated that there was a need to integrate both pendulum and geodetic data in dam monitoring.

Gamse et al (2016) studied the behaviour of the Alqueva Dam to understand the reversible displacements of the dam measured by both geodetic methods and pendulums. Data spanning for 13 years was used in developing an HST model for radial and tangential displacements. Results showed that there was a difference of 8mm in radio and 1mm in tangential direction between the displacements measured by geodetic methods and pendulums.

Wang et al., (2019) developed a new model called hydraulic hysteretic season and time (HHST) model to predict the deformation of Jinping I Arch Dam in China. This model was proposed to detect the abnormal behaviour of the dam that was caused by the hysteretic hydraulic deformation and ambient temperature drop effect. The length of the data used was 5 years and results from the model showed that a 70% change in deformation was due to ambient temperature drop and 30% was due to viscoelastic hysteretic hydraulic deformation.

Li and Wang (2019) compared the performance of two statistical models namely: - multiple linear regression and artificial neural network in the modelling of measured deformations from Dongjang Dam located in China. Data used in the development of the models spanned from February 2003 to December 2013. Results showed that artificial neural networks performed better than multiple linear regression especially with the non-linear effects and interactions between the different variables.

3.2.2.7 Partial least squares regression

Partial least square regression (PLSR) is a regression method developed for multiple linear regression based on the decomposition of matrices of response variables and predictor variables. PLSR analyses data with strong collinear, noisy and numerous response variables Y (Wold et al., 2001). This technique identifies a linear regression model by projecting the predicted variables and the response variables into a new lower-dimensional space to control for collinearity among the variables (Tobias, 1995; Van Roon et al., 2014).

In this methodology, response variables are predicted by finding latent variables (T), that mode X and simultaneously predict Y. Equation 3.11 shows the generalised PLSR model expressed as a double decomposition of X and predicted $\hat{Y}$ (Abdi and Williams, 2013).

$$X = TP^T$$
$$\hat{Y} = TBC^T \tag{3.11}$$

Where P and C are the X and Y loadings on X and Y respectively and B is a diagonal matrix. These latent variables are ordered according to the amount of $\hat{Y}$ that they explain. The latent variables are computed by singular value decomposition iteratively. Each procedure of the SVD produces orthogonal latent variables for X and Y and corresponding regression weights.

Let X and Y be mean centred and normalized such that the mean of each column is zero and its sum of squares is one. Initially, X and Y are stored (respectively) in matrices X_0 and Y_0. The matrix of correlations between X_0 and Y_0 is calculated as:

$$Z_1 = X_0^T Y_0 \tag{3.12}$$

Performing singular value decomposition on Z_1 gives two sets of orthogonal singular vectors W_1 and C_1, and the corresponding singular values Δ_1.

$$Z_1 = W_1 \Delta_1 C_1^T \tag{3.13}$$

The first pair of singular vectors (i.e., the first columns of W_1 and C_1) are denoted w_1 and c_1 and the first singular value is denoted δ_1. The singular value represents the maximum covariance between the singular vectors. The first latent variable of X is given by:

$$t_1 = X_0 w_1 \tag{3.14}$$

where t_1 is normalized. The loading of X_0 on t_1 is given by:

$$p_1 = X_0^T t_1 \tag{3.15}$$

The least-square estimate of X from the first latent variable is given by

$$\hat{X}_1 = t_1^T p_1 \tag{3.16}$$

Similarly, the first latent variable for Y denoted by u_1 is calculated as:

$$u_1 = Y_0 c_1 \quad (3.17)$$

Y is then reconstituted from its latent variable as:

$$\hat{Y}_1 = u_1 c_1^T \quad (3.18)$$

The Equation 3.18 above can be rewritten to obtain the prediction Y from the X latent variable as

$$\hat{Y}_1 = t_1 b_1 c_1^T \quad (3.19)$$

With

$$b_1 = t_1^T u_1 \quad (3.20)$$

The scale b_1 is the slope of the regression of $\hat{Y}_1$on t_1. Matrices $\hat{X}_1$ and $\hat{Y}_1$ are then subjected from the original X_0 and Y_0 to give a deflated X_1 and Y_1, i.e.,

$$X_1 = X_0 - \hat{X}_1 \text{ and } Y_1 = Y_0 - \hat{Y}_1 \quad (3.21)$$

The process of iteration continues until X is completely decomposed into L components ($L \leq rank\ (X)$. The weights w_l for X are stored in matrix W, and the latent variables are stored in matrix T. Matrix C stores the weights for Y while U stores the pseudo latent variables of Y. The loadings for X are stored in matrix P and the regression weights are stored in the diagonal matrix B. These regression weights are used to predict Y from X.

To understand the effect of water level and temperature on the horizontal displacements of the Gezhouba Dam located in China, Wenbo and Weibin (2014) used the least square method and partial least squares regression method. Two methods namely non-floating tensile wire and the vacuum laser collimation system were used in the measurement of displacements. Results showed that the two systems performed well with the temperature being factors that affect the fluctuation of horizontal displacement of the dam with 90% effect and 10% for water level component.

3.2.2.8 Independent component regression

Independent component regression (ICR) is a method that extracts mutually independent components from arguments first and then builds a regression model with independent components instead of the observed factors (Shao et al., 2006). Dai et al., (2013) summarized ICR as follows:

(1) Extract the independent components C of the observed factors X through a FastICA algorithm

$$\begin{cases} [A, W] = FastICA(X) \\ C_{nxl} = X_{nxm} W^T \end{cases} \quad (3.22)$$

(2) Build a regression model with C and Y and calculate the coefficients of regression using the least-squares method.

$$\begin{cases} Y = d_0 + d_1 C_1 + d_2 C_2 + \cdots d_l C_l \\ D = [d_0, d_1 \ldots d_l]^T \end{cases} \quad (3.23)$$

(3) Calculate the original coefficients of regression with the result in D (Eequation 3.23) and the separating matrix W in Equation (3.22)

ICR has two advantages namely (Dai et al 2013):-

(i) The components are mutually independent, so the problem of ill condition on the coefficient matrix is avoided.

(ii) The ICs extracted from ICR can be used to identify the implicit factors relative to the dependent variable Y.

Popescu and Manolescu (2008) utilised independent component analysis in separating hydrostatic, temperature and time effects from displacements measured from Vidraru Dam located in Romania. Popescu and Manolescu (2008) used the so-called Second Order Blind Identification (SOBI) algorithm to analyse on 1200-day data. Results indicated that ICA can be used to separate signals into hydrostatic load, temperature and time effects.

Dai et al (2013) modelled and identified the physical origins of deformation of Wuqiangxi Dam in China. The authors used independent component regression in the analysis of data such that ill condition problem in the coefficient matrix could be prevented. Horizontal displacements, daily temperature and water level were used in the modelling. The algorithm was compared with principal component regression and results show that it performed better than the latter.

3.2.2.9 Multivariate Adaptive Regression Splines

The Multivariate Adaptive Regression Splines (MARS) were introduced for fitting the relationship between a set of predictors and dependent variables (Friedman 1991). MARS is a multivariate, piecewise regression technique that can be used to model complex relationship. The space of predictors is divided into multiple knots to fit a spline function between these knots (Friedman 1991). The MARS algorithm searches over all possible univariate hinge locations and across interactions among all variables. It does so through the use of combinations of variables called basis functions.

MARS uses expansions in piecewise linear basis functions of the form $(x-t)_+$ and$(t-x)_+$. This leads to truncated functions in the form of :

$$(x-t)_+ = \begin{Bmatrix} x-t, & if\ x>t, \\ 0, otherwise & \end{Bmatrix} \text{ and } (t-x)_+ = \begin{Bmatrix} t-x, & if\ x<t, \\ 0, otherwise & \end{Bmatrix} \quad (3.24)$$

Figure 3.1 below shows an example of two basis functions

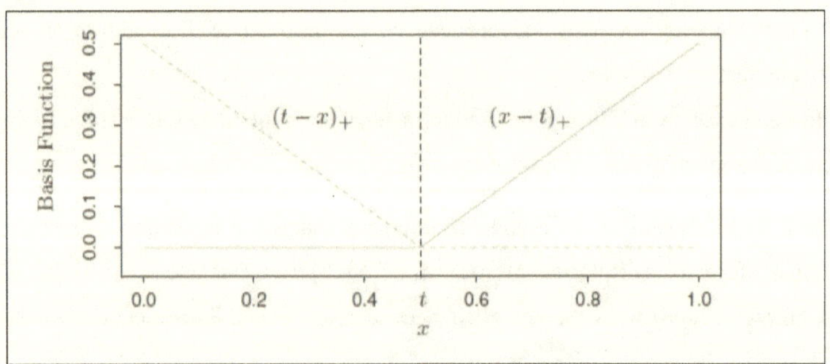

Figure 3.1. Basis functions (Hastie et al., 2009)

Each function is piecewise linear, with a *knot* at the value *t* and these are linear splines. These two functions are called a *reflected pair*. The idea is to form reflected pairs for each input *Xj* with knots at each observed value *xij* of that input. Therefore, the collection of basis functions is:

$$C = \{(X_j - t)_+, (t - X_j)_+\} \quad (3.25)$$

Where $t \in \{x_{1j}, x_{2j}, \dots \dots, x_{Nj}$
$j = 1,2,3, \dots \dots \dots \dots \dots . p$

The model-building strategy is like a forward stepwise linear regression, but instead of using the original inputs, we are allowed to use functions from the set *C* and their products. Thus the model has the form

$$f(X) = \beta_0 + \sum_{m=1}^{M} \beta_m h_m(X) \tag{3.26}$$

Where $h_m(X)$ is the spline function in C, β_0 to β_m are the coefficients of the model that estimates to yield the best fit to the data, and M is the number of sub-regions or the number of basis functions in the model.

This model searches over the space of all inputs and predictor values (referred to as "knots") as well as the interactions between variables. During this search, an increasingly larger number of basis functions are added to the model to minimize a lack-of-fit criterion. As a result of these operations, MARS automatically determines the most important independent variables as well as the most significant interactions among them. It is noted that the search for the best predictor and knot location is performed in an iterative process. The predictors as well as the knot location, having the most contribution to the model, are selected first. Also, at the end of each iteration, the introduction of interaction is checked for possible model improvements.

The MARS algorithm searches for all possible knot locations for each variable. The forward stepwise addition procedure can produce a large collection of basis functions, and the process is stopped when a user-specified maximum model size is reached. Then a backward pruning procedure is applied to remove one at a time any non-constant basis functions that no longer make sufficient contribution to the model. The best-fitting model in the stepwise sequence is chosen to minimize the generalized cross-validation criterion of Craven and Wahba (2001): This is defined as:

$$GCV = \frac{1}{N} \frac{\sum_{i=1}^{N} (y_i - \hat{f}(X_i))^2}{\left[1 - \frac{\tilde{C}(M)}{N}\right]^2} \tag{3.27}$$

Where $\left[1-\frac{\check{C}(M)}{N}\right]^2$ a complex is function and $\check{C}(M)$ is the effective number of parameters in the model. This accounts both for the number of terms in the models, plus the number of parameters used in selecting the optimal positions of the knots.

This is defined as:

$$\check{C}(M) = C(M) + dM \tag{3.28}$$

Where C (M) is the number of parameters being fit and d represents a cost parameter of the procedure. The higher the cost d is, the more basis functions will be eliminated (Put et al., 2004).

MARS has not been widely used in analysing data from dam monitoring systems; below is one case study where MARS has been used.

Salazar et al., (2015) compared the prediction capability of different algorithms in the modelling of displacement and leakage of La Pelle Dam. These included random forest, neural network, support vector machines, multivariate adaptive regression splines and boosted regression trees. Using the mean absolute error and average relative variance, the accuracy of the models was determined. Results showed that the boosted regression trees performed better than the other algorithms.

In summary, regression-based techniques have been widely used in dam monitoring because they are easy to use. However, these techniques have several drawbacks such as (Salazar et al., 2015):-

(i) Functions have to be defined beforehand and this does not represent the true behaviour of the dam structure.

(ii) The governing variables are supposed to be independent, although some of them have been proven to be correlated.

(iii) Not well-suited to model non-linear interactions between input variables.

3.2.3 Machine learning-based models

Machine learning models that adjust their parameters to perform better when exposed to more data, i.e. they adapt to the given data to perform a certain objective. In dam monitoring, machine learning algorithms have been utilised to predict the behaviour of dams especially in cases of non-linearity where regression-based models mentioned above performed poorly.

3.2.3.1 Support vector regression

The concept of support vector regression (SVR) is an extension of support vector machines which were developed by Vapnik (1995). SVR proposed by Drucker et al. (1997) involves approximation of the regression function within a so-called 'tube' of margin ε using a set of support vectors that belong to the training data.

Suppose there exists training data $\{(x_1, y_1), \dots \dots, (x_n, y_n)\}$ where $x_i \epsilon R^n$ is the input data and $y_i \epsilon R^n$ is the output data available for building a regression model. SVR algorithm aims to identify a regression function $y = f(x)$ in a transformed feature space F that predicts accurately that the outputs correspond to a new set of inputs. The function f(x) is known as the hard ε-band hyperplane of a given sample set shown in Figure 3.2 below.

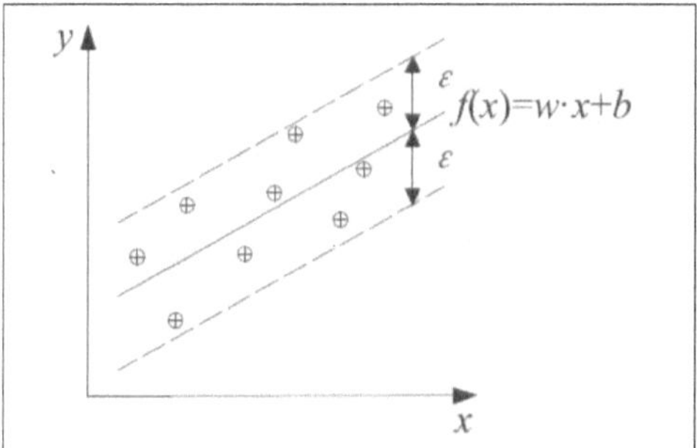

Figure 3.2. Hard ε-band hyperplane (Smola and Scholkopf, 2004)

The SVR algorithm applies a nonlinear function Φ to the original data from the initial input space to a higher dimensional feature space F:

$$\phi: R^n \to F, \omega \epsilon F \tag{3.29}$$

$$f(x) = \langle \omega \Phi(x) \rangle + b \tag{3.30}$$

Where ω is the weight vector and b is a constant that relates to the offset of the function. SVR performs linear regression in the higher-dimensional feature space by the ε-insensitive loss function. The aim is to find a function that fits the training data with a deviation less or equal to ε and at the same time is as flat as possible. This means finding a small weight vector ω by minimising the quadratic norm of the vector ω (Smola and Scholkopf, 1998). To allow for outliers, ξ_i, ξ_i^* are positive variables that measure the error at the upper and the lower sides of the tube. These variables are introduced leading to (Vapnik, 1995):

$$Min\frac{1}{2}\|\omega\|^2 + C\sum_{i=1}^{l}(\xi_i + \xi_i^*) \quad (3.30)$$

$$\text{Subjected to } y_i - \langle\omega, \Phi(x_i)\rangle - b \leq \varepsilon + \xi_i$$

$$\langle\omega, \Phi(x_i)\rangle + b - y_i \leq \varepsilon + \xi_i^*$$

$$\xi_i, \xi_i^* \geq 0, i = 1,2,3 \ldots\ldots, l \quad (3.31)$$

Parameter C measures the trade-off between generalization ability and accuracy in the training data and parameter ε defines the degree of tolerance to errors. The above problem stated in Equation 3.30. can be solved by presenting it in a dual form. That is to say, constructing a Lagrange function and then applying saddle point conditions. This leads to a solution as suggested by (Vapnik, 1998):

$$\omega = \sum_{i=1}^{l}(\alpha_i - \alpha_i^*)\Phi(x_i) \quad (3.32)$$

$$f(x) = \sum_{i=1}^{l}(\alpha_i - \alpha_i^*)\Phi(x_i)\, K(x_i, x) + b \quad (3.33)$$

Where α_i *and* α_i^* are Lagrange multipliers and $K(x_i, x)$ is defined as a kernel function (Vapnik, 1998). The function allows us to obtain a solution for the original regression problem, without considering the transformation $\Phi(x)$ applied to the data.

There are four commonly used kernels namely linear kernel, polynomial kernel, radial basis function (RBF) or Gaussian kernel and the sigmoid kernel. The type of kernel function influences the parameters of the SVR kernel. This is because if the kernel function and parameters are not chosen properly it can affect the regression accuracy. The expressions of these kernels are shown below:

Linear kernel $K(x_i, x_j) = x_i^T x_j$

Polynomial kernel $K(x_i, x_j) = (1 + x_i . x_j)^d$

RBF kernel $K(x_i, x_j) = \exp(-\gamma\|x_i - x_j\|^2)$

Sigmoid kernel $K(x_i, x_j) = tanh[v(x_i, x_j) + \alpha]$

The RBF is the most commonly used kernel function as it has advantages of being effective and has fast training process (Xin et al., 2012). When using the RBF kernel function, the following parameters need to be determined (Smola and Scholkopf, 2004)

(i) Regularization parameter C which is used in determining the trade-off cost between minimising training error and minimising model complexity
(ii) Kernel parameter γ that represents the RBF kernel function
(iii) The tube size of the ε-insensitive loss function

Cheng and Zheng (2013) presented results from two multivariate dam safety models subjected to horizontal displacements and hydraulic uplift pressure of Mianhuatan Gravity Dam. These models include:- (1) upper control limits which were calculated by performing kernel density function on the square prediction error and (2) the least square support vector machine adapted to simulate nonlinear mapping from environmental variables to latent variables. These are predicted and the prediction interval is calculated to provide control range for future monitoring data.

Su et al (2015) developed two monitoring displacement models namely static model and real-time updated model based on support vector machines. The study aimed to understand the adaptability of the support vector machine models in dam safety monitoring and also reducing the modelling time using horizontal displacements for four years on a gravity dam. The study showed that the choice of the insensitive loss function and penalty factor is key while using support vector machines in nonlinear regression problems.

Su et al (2016) extended the support vector machine methodology by introducing the so-called chaos theory and particle swarm optimization which was combined in the prediction of dam deformation. The displacements of the crest of a dam located in China were measured using pendulums and the data used in the modelling spanned from 2003 to 2007. Results indicated that the proposed methodology performed better than the SVM in the forecasting of displacements.

Kang et al (2016) predicted horizontal displacements of Fengman Dam located in China based on least squares support vector machine. The length of the data used in the analysis spanned from 1997 to 2009. Results obtained from the proposed methodology were compared with the multiple linear regression and step regression models and authors concluded that the SVR model offered a higher prediction accuracy and the training speed was faster.

3.2.3.2 Neural Networks

Artificial Neural Networks (ANN) are in a group of artificial intelligence that has been used widely in research areas of engineering and science (Flood et al, 1994). Awodele and Owalale (2009) described ANN as an information processing paradigm that is inspired by the way biological nervous systems, such as the brain, process information. This system simulates the ability of a biological neural network by interconnecting many simple neurons as shown in the figure below (Jeng et al., 2003). According to Jeng et al (2003), the following characteristics of ANN enable them to model so many problems in engineering: - (i) ability to learn (ii) memory distribution (iii) operating in parallel and (iv) fault tolerate. Figure 3.3 shows the typical structure of an artificial neural network.

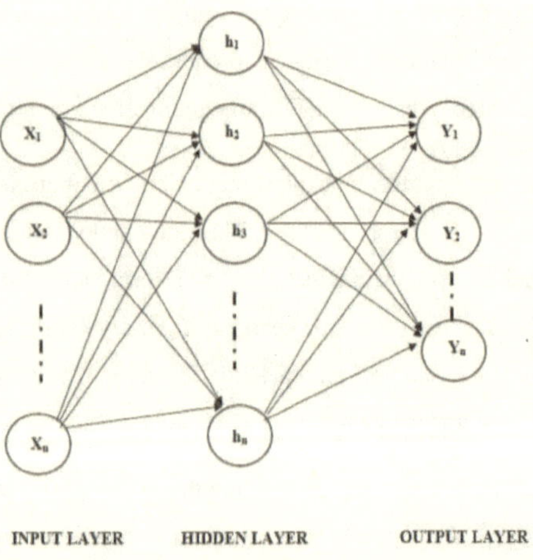

Figure 3.3. The architecture of the ANN structure

The units in the middle of the network, computing the derived features hm, are called hidden units because the values hm are not directly observed while X and Y are the so-called input layer and output layer.

The artificial neuron accepts input signals, processes it through some calculations and then produces output signals. The network consists of several connected neurons grouped in layers. The number of layers in a network determines its complexity. The input layer is fed by the input signal which is then processed by the hidden layer and comes out through the output layer.

The network also contains weight coefficients which express the importance of each neuron's input and determines the input's capability for simulation of the neurons (Flood, 1994). Each weight coefficient is assigned to each input neuron and by adding the product of the weight coefficients and input signal the input signal from each neuron is calculated.

The transformation function determines whether the result from the input signal can generate an output. Several transfer functions are used for this purpose. These functions include (Araghinejad, 2014):-

- Linear function

$$f(x) = x$$

- Symmetric-saturating-linear

$$f(x) = \begin{cases} \delta & x \geq \theta \\ x & -\theta \leq x \leq \theta \\ -\delta & x \leq -\theta \end{cases}$$

- Log sigmoid

$$f(x) = \frac{1}{1 + e^{-\alpha x}} \alpha > 0$$

- Tangent sigmoid

$$f(x) = \left(\frac{2}{1+e^{-\alpha x}}\right) - 1 \quad \alpha > 0$$

- Radial basis

$$f(x) = e^{-x^2/\sigma^2}$$

Neural network architecture

Haykins (1999) divided NN architecture into three classes namely:

- Single-layer feedforward networks

In this type of network, the input player of source nodes only projects onto the output layer of neurons in one direction as shown in Figure 3.4 .

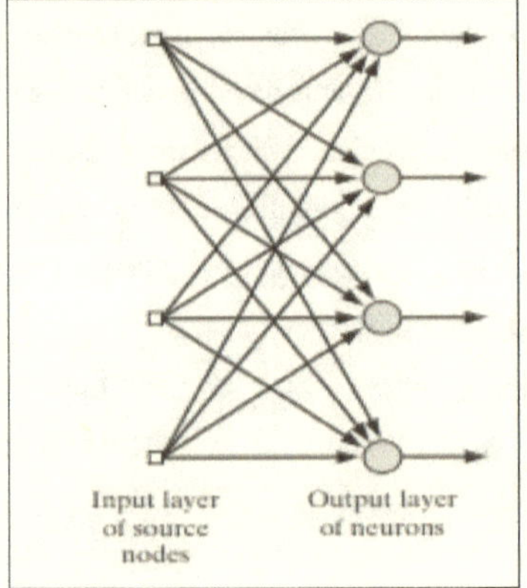

Figure 3.4. Feedforward network with a single layer of neurons

- Multilayer feedforward networks

This type of feedforward network is different from the single-layer due to the presence of more than one hidden layers. The presence of these hidden layers is to enable the network to extract higher-order statistics (Haykins, 1999). Figure 3.5 shows a typical multilayer feedforward network.

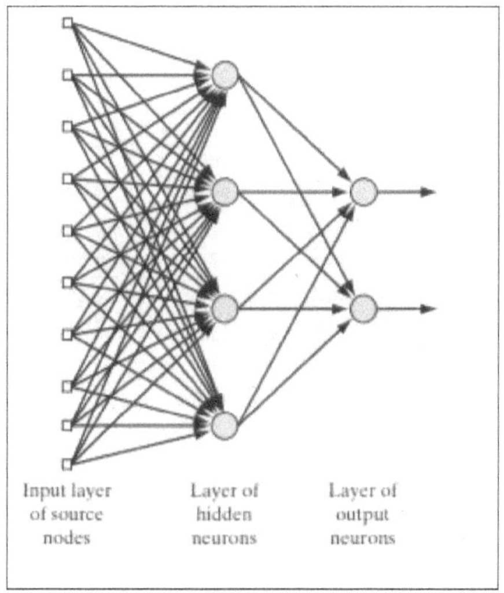

Figure 3.5. Multilayer feedforward network (Haykins, 1999)

- Recurrent networks

Recurrent networks are different from the feedforward neural network in such a way that they contain at least one feedback loop (Haykins, 1999) as shown in Figure 3.6 below. In this type of network, loops are introduced due to signals moving in both directions.

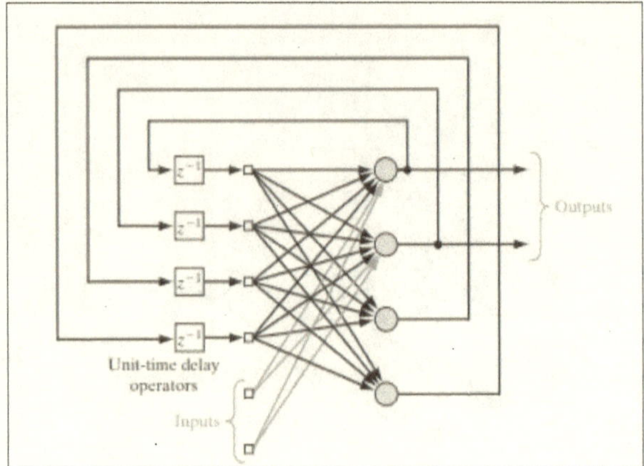

Figure 3.6. Recurrent network (Haykins, 1999)

Training networks

The most important part of neural networks is the training process which includes the selection of suitable weighting sets that make the network perform to its intended use.

- Learning process

This process consists of periodic data transmission through the network and compartment of the received input values with the expected ones (Lazarevska et al., 2014). This can be divided into general algorithms (Awodele and Jegede, 2009).

(i) Associative mapping: In this process, the network learns to produce a particular pattern on the set of input signals whenever another particular pattern is applied to the set of inputs.

(ii) Auto-association: in this process, an input pattern is associated with itself and the states of inputs and outputs coincide.

Yinghua and Chang (2010) developed a displacement model based on three stages namely PCA, BP-ANN and deformation forecast. This was done in conjunction with ANN and genetic algorithms (GA). This procedure improved the performance of pure BP-ANN in terms of convergence ability. However, it is limited to displacements contributed by water level, temperature and ageing variability.

Mata (2011) compared two statistical models for the characterization of Alto Rabago Dam behaviour under environmental loads of reservoir level and external temperature of structural responses (crest displacements). These models include multiple linear regression (MLR) and neural network (NN) models displacements at every point of the dam were strongly linked to the corresponding variation in water level in the reservoir. These observations were used to develop and calibrate MLR and NN models based on experimental data time histories (over 25 years) of reservoir level and external temperatures and of crest displacements. Mata (2011) concluded that the NN model performed better in months with extreme temperatures than the MLR model.

Loh et al. (2011) established advanced statistical methods to extract trends from long term structural health monitoring data and attempt to set an early warning threshold level based on the results of the analyses of Fei-Tsui Dam located in Taiwan. Data on daily deformation of the dam at a single sample per day and the temperature data were measured for 22 years. Data was analysed using the singular spectrum analysis with the autoregressive model (SSA-AR) and the nonlinear principal component analysis (NPCA) using auto-associative neural network method (AANN). Results indicated that the AANN method could capture the periodic variations (temperature, season) as well as the trends (creep). This made the AANN model suitable for a long-term trend prediction over the SSA-AR model which can only be used for short-term prediction.

Hu et al., (2011) assessed the performance of different statistical models namely: - regression model, statistical model, backpropagation neural network model and the neural network merging model in the analysis of dam deformation monitoring data. Vertical displacements collected at an observation point at CC Dam used in the analysis ranged between January 1999 and December 2006. The comparison of the models was by Root Mean Square Error. Results showed that the neural network merging model performed better than the other models with a 25% increase in the prediction accuracy.

Demirkaya and Balcilar (2012) compared two statistical methods namely multiple linear regression (MLR) and multiple-layer perceptron (MLP) models to build a daily displacement forecasting system for the Schlegeis Arch Dam. The independent variables used in the development of the statistical models were the time histories water level, air temperature and concrete temperatures. These were measured at six points on the dam with one value per day for 8 years. Demirkaya and Balcilar (2012) concluded that MLR models performed better than MLP's under the criteria of R^2. Results suggested that linear regression offered the most effective solution of linear problems and are suitable for analysing data collected from static monitoring of dams.

Kao and Loh (2013) compared the performance of ANN-based approaches namely:- SNN, NARXNN, PCA and ANN with a purpose of understanding the behaviour of Fei-Tsui Dam using deformation data. Using residual deformation, threshold levels were calculated using statistical analyses. The authors concluded that SNN, NARXNN and ANN performed well in the modelling of deformation of the dam. This was exhibited with the high values of R^2 which were close to 1.

Kang et al (2017) presented a model based on an extreme learning machine to predict the deformation of Fengman Concrete Gravity Dam located in Jilin City, China. Ambient temperature and water level data sets were used as predictors to predict the deformations of the dam. The data used in the analysis spanned over 10 years. The developed models were compared with the multiple linear regression, stepwise regression and backpropagation models and results showed that the extreme learning machine model gave more accurate results than the other models and were faster than the BP model.

Wei et al., (2019) developed a combination forecast model using residual correction for displacement data of a concrete dam in China. Two years of monitoring data were used to construct the three models namely: HST, BP and the combined model. Residuals from the HST model were used to reconstruct the new model based on BP network and ARIMA. Results showed that the proposed model reduced the forecast error as compared to the traditional models.

Artificial neural networks have been widely used in dam monitoring because they can deal with the nonlinear interactions between the loads and dam responses. However, ANN has some drawbacks. These include (Salazar et al., 2015): -

(i) Results depend on the initialisation of the weights

(ii) The best network architecture i.e. the number of hidden layers and neurons in each layer is not known beforehand.

(iii) The training process may reach a local minimum of the error function

3.2.3.3 Random forests

Random forests are kind of integrated learning algorithm which combines the performance of different decision tree algorithms used to predict a variable (Breiman, 2001). This method employs the strategy of a random selection of a subset of m predictors to grow binary trees, where each tree is grown on a bootstrap sample of the training set as shown in Figure 3.7 below. Random forest algorithm receives an input vector (x) made of values of different features analysed for a given training area and then builds a number K of regression trees and averages results (Rodriguez-Galiano et al., 2015). This results in $\{\{T(x)\}\}_1^K$ trees that have grown and the regression predictor becomes:

$$f_{rf}^{K}(x) = \frac{1}{K}\sum_{K=1}^{K} T(x) \qquad (3.34)$$

To avoid the correlation of the different trees, random forest increases the diversity of the trees by making them grow from different training data subsets. This process is known as bagging, i.e. method used to create training data by resampling randomly the original dataset with replacement, i.e., with no deletion of the data selected from the input sample for the generation of the next subset (Rodriguez-Galiano et al., 2015). According to Peters et al., (2007), the samples that are not selected in the kth tree during the process of bagging are included as part of a subset called out-of-lag (oob) which can later be used for performance evaluation.

In random forest modelling, three parameters have to be satisfied (Li et al., 2016):-

(1) The number of trees to grow in the forest (n_{tree})

(2) The number of randomly selected predictor values at each node (m_{try}) and

(3) The minimal number of observations at the terminal nodes of the trees (node sizes)

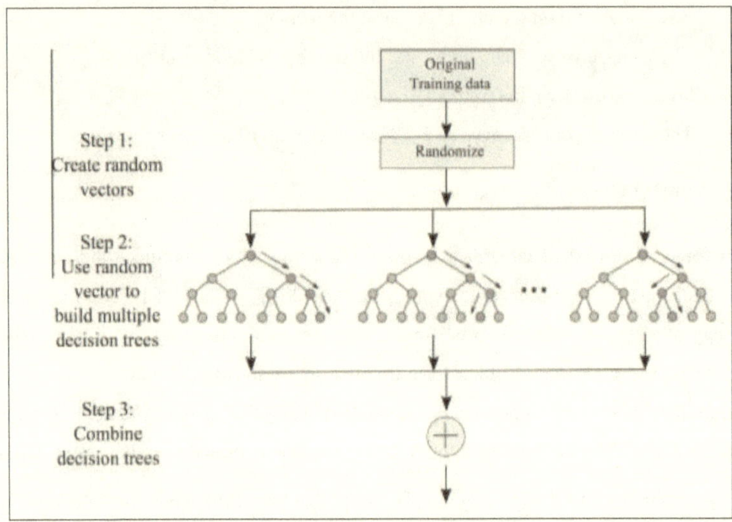

Figure 3.7. Random forest methodology (Malekipirbazari and Aksakalli, 2005)

Random forest methodology offers the following advantages (Rodriguez-Galiano et al., 2015):-

(1) It runs efficiently on large databases

(2) It can handle thousands of input variables without variable deletion

(3) It is relatively robust to outliers and noise

(4) It is computationally lighter than other ensemble methods for example boosting.

Dai et al., (2018) predicted the horizontal deformations of a concrete gravity dam using random forest regression. Data used in the development of the statistical model spanned from January 2003 to December 2008 and was measured daily. Results showed that random forest regression can be used in the prediction of the structural behaviour of dams because of its high accuracy and low error alarm.

3.2.3.4 Boosted regression trees

Boosted regression trees (BRT) is one of several techniques that aim to improve the performance of a single model by fitting many models and combining them for prediction. BRT uses two algorithms: regression trees are from the classification and regression tree (decision tree) group of models, and boosting builds and combines a collection of models (Elith et al., 2008). Regression trees are based on the recursive division of the training space into disjointed regions (Salazar et al., 2017).

Boosting is a forward and stage-wise procedure in which a subset of the data is randomly selected to iteratively fit new tree models to minimize the loss function (Elith et al., 2008). This process introduces a stochastic gradient boosting procedure that can improve model performance and reduce the risk of overfitting (Friedman, 2002). The BRT algorithm is an iterative process in which tree-based models were fitted iteratively using recursive binary splits to identify poorly modelled observations in existing trees until minimum model deviance was reached. The final fitted model is a linear function of the sum of all trees multiplied by the learning rate (LR) based on all data (Elith et al., 2008). In BRT modelling, four parameters are user-defined: the learning rate (LR), tree complexity (TC), number of trees (NT) and bag fraction (BF). LR represents the contribution of each tree to the final fitted model, and TC controls the size of trees and whether interactions between variables should be considered.

Elith et al. (2008) explained BRT as follows. The initial regression tree is the one that reduces the loss function the most. At each iteration, the focus is on the residuals and root mean square error reduction. In the second step, a regression tree, which can contain different variables and split points than the first tree, is fit to the prediction residuals of the first tree. The overall model now contains two trees (i.e., two terms), and the residuals from this two-term model are estimated. The process is stage-wise, i.e., existing trees are left unchanged as the model grows increasingly larger. Only the fitted value for each observation is re-estimated at each step to reflect the contribution of the newly added tree. In the end, the final BRT model is a linear combination of 22 numerous trees and can be thought of as a regression model with each term being a tree.

The advantages of boosted regression trees include (Salazar et al., 2017): -

(i) being robust
(ii) requires little data pre-processing
(iii) works well with non-linear relationships and can interact well among predictors.

However, they are unstable i.e., small variations in the training data which may result in big variations (Elith et al., 2008).

Salazar et al (2015) carried out a comparative study on the performance of three analysis methods namely: - (i) HST, (ii) RF and (iii) BST using monitoring data recorded from Itaipu Dam located in Brazil. The authors reported that the machine learning algorithms offered accurate results for radial and tangential displacements as compared to HST.

3.2.4 Time series-based models

3.2.4.1 Autoregressive (AR) models

Autoregressive models are a type of time series model that utilise the past observed values in predicting the present and future values. The resulting relationship is presented in Equation 3.35 below (Gombay and Seban, 2009):

$$Y_i - \mu = \phi_1(Y_{i-1} - \mu) + \cdots + \phi_p(Y_{i-p} - \mu) + \varepsilon_{i,} \; i \geq 1 \qquad (3.35)$$

where μ is the mean, ϕ_i is …… and ε_i is the error with mean zero and variance σ^2

In AR modelling, it is important to choose the right model insensitive and this can be obtained through autocorrelation (ACF) and partial autocorrelation functions (PACF) calculation. Autocorrelation measures the linear dependence of a variable at different lags while partial autocorrelation measures the linear dependence between different lags after removing the mutual autocorrelation in-between the lag (Shumway and Stoffer, 2010). Regression trees have a disadvantage of not being able to detect gradual anomalies (Yao et al., 2010)

3.2.4.2 Nonlinear Autoregressive with eXogenous inputs (NARX) model

The Nonlinear Autoregressive with Exogenous inputs (NARX) model is characterised by the non-linear relations between the past inputs, past outputs and the predicted process output and can be delineated by the high order difference equation (Lin et al, 1996, Leontaritis and Billings 1985a):-

$$y(t) = f\{y(t-1), y(t-2), \ldots\ldots, y(t-n_y), u(t-1), u(t-2), \ldots\ldots, u(t-n_u)\} + \varepsilon(t) \qquad (3.36)$$

where $u(t)$ and $y(t)$ are the model input and output at time t respectively. The current output $y(t)$ depends entirely on the current input $u(t)$. n_u and n_y are in the input and output orders of the dynamic model also known as maximum lags and $\varepsilon(t)$ is the error or white noise term. The function f is nonlinear and can be approximated by the regression model of the form:

$$y(t) = \sum_{i=0}^{n_u} a(i).u(t-i) + \sum_{j=1}^{n_y} b(i).y(t-j) + \sum_{i=0}^{n_u}\sum_{j=i}^{n_u} a(i,j).u(t-i).u(t-j) + \sum_{i=0}^{n_u}\sum_{j=i}^{n_y} c(i,j).u(t-i).y(t-j) + \varepsilon(t) \quad (3.37)$$

Where $a(i)$ and $a(i,j)$ are the coefficients of linear and nonlinear for originating exogenous terms; $b(i)$ and $b(i,j)$ are the coefficients of the linear and nonlinear autoregressive terms; $c(i,j)$ are the coefficients of the nonlinear cross terms. Equation 3.36 can also be expressed as

$$y(t) = \varphi^T(t-1)v + \varepsilon(t) \quad (3.38)$$

Where $\varphi^T(t-1)$ is the regressor vector containing linear and nonlinear combinations of the input and output variables and v is the parameter vector containing the polynomial coefficients.

The selection of model order: the important step in estimating NARX models is to choose the model order. The model performance is evaluated by the Means Squared Error (MSE) and the Sum Squared Error (SSE).

Model validation: The model is validated with two sets of validation data which are unseen independent data sets that are not used in NARX model parameter estimation.

Piroddi and Spinelli (2003) predicted radial crest displacement of Schlegesis Dam using an identification algorithm for polynomial NARX model. This algorithm combines a model selection procedure based on the minimization of the simulation error and pruning mechanism for the elimination of redundant terms. Results showed that this method performed better than the classical HST.

Rankovic et al (2014) applied support vector regression based on Non-linear Auto-Regressive with eXogeous outputs (NARX) model to accurately and forecast tangential displacements of Djerdap Hydropower Plant. Water levels both upstream and downstream and tangential displacements at previous steps were used as inputs. The previous tangential displacements were used to capture the thermal loads. Prediction results were in agreement with the experimental data with a Pearson coefficient of 0.9940. It was also concluded that support vector regression is an effective tool for approximation of nonlinear behaviour of dams.

Radovanovic et al (2015) presented a methodology based on NARX for statistical modelling if monitoring data of Vrutic Concrete Gravity Dam located in Serbia namely. Radial displacements of the dam were used in the analysis with water level and air temperature ranging from 1995-2002. The results were compared with those from the MLR and it was seen that the proposed model performed better than the MLR.

Garcia et al (2015) developed a model to forecast horizontal displacements of a block on Itaipu Dam in Brazil. Two approaches namely:- autoregressive distributed lags and bound testing were used to relate water level, air temperature and horizontal displacements for data ranging from the year 2000 to 2015. It was concluded that the presented models can be used to confirm relationships and can also be used if the time series is stationary.

Oltean et al.,(2017) proposed a multi-step model to predict the displacement of a dam using autoregressive neural network. Previous values of horizontal displacements, water level and temperature recorded for 456 months were used in the development of models. Results showed that the proposed methodology performed well with an R^2 value greater than 9 and it could also detect abnormalities.

3.2.4.3 Adaptive Neuro-Fuzzy Systems (ANFIS)

Adaptive neuro-fuzzy inference system (ANFIS) is a kind of neural network that is based on the Sugeno fuzzy inference system (Araghinejad, 2014). ANFIS is a class of adaptive networks that incorporate both neural networks and fuzzy logic principles. (Mathur et al., 2016). The use of ANFIS can make the selection of the rule base more adaptive to the situation. In this technique, the rule base is selected utilizing the neural network techniques via the back-propagation algorithm. To enhance its applicability and performance, the properties of fuzzy logic, that is, approximating a non-linear system by setting IF-THEN rules are inherited in this modelling technique (Mathur et al., 2016).

A conceptual ANFIS consists of five components: inputs and output database, a Fuzzy system generator, a Fuzzy Inference System (FIS), and an Adaptive Neural Network (Kumar and Kumar, 2012).

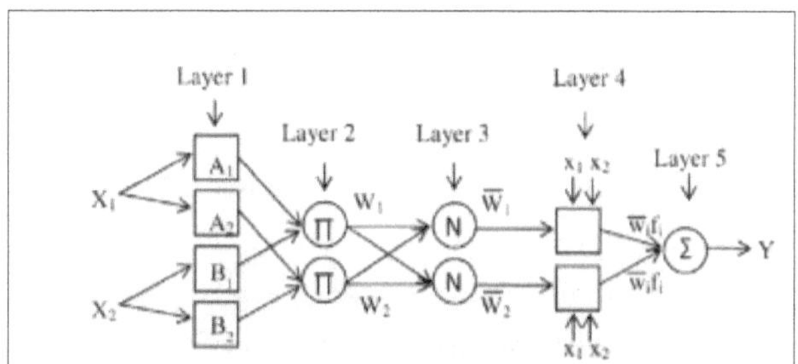

Figure 3.8.ANFIS architecture (Jang et al., 1997)

ANFIS is a simple data learning technique that uses a fuzzy inference system model to transform a given input into a target output. This prediction involves membership functions, fuzzy logic operators and if-then rules. There are two types of fuzzy system, commonly known as the Mamdani and Sugeno models. There are five main processing stages in the ANFIS operation, including input fuzzification, application of fuzzy operators, application method, output aggregation, and defuzzification.

Architecture of ANFIS

The ANFIS is a multilayer feedforward network which uses neural network learning algorithms and fuzzy reasoning to map an input space to an output space. With the ability to combine the verbal power of a fuzzy system with the numeric power of a neural system adaptive network, ANFIS is powerful in modelling several systems. This is because ANFIS possesses the good capability of learning, constructing, expensing, and classifying. It has the advantage of allowing the extraction of fuzzy rules from numerical data or expert knowledge and adaptively constructs a rule base. Furthermore, it can tune the complicated conversion of human intelligence to fuzzy systems. The main drawback of the ANFIS predicting model is the time requested for training structure and determining parameters, which took much time. For simplicity, we assume the fuzzy inference system under consideration has two inputs, x and y, and one output, z. For a first-order Sugeno fuzzy model (Takagi and Sungeno, 1985), a typical rule set with two fuzzy if-then rules can be expressed as:

Rule 1: If x is A_1 and y is B_1 then $w_1 = u_1x + v_1y + r_1$

Rule 2: If x is A_2 and y is B_2 then $w_2 = u_2 x + v_2 y + r_2$ (3.39)

where u_i, v_i and r_i (i = 1 or 2) are linear parameters in the then-part (consequent part) of the first-order Sugeno fuzzy model. The architecture of ANFIS consists of five layers (Figure 3.8), and a brief introduction of the model is as follows.

Layer 1: input nodes. Each node of this layer generates membership grades to which they belong to each of the appropriate fuzzy sets using membership functions (MFs)

$$O_{1,i} = \mu_{A_i}(x) \text{ for } i = 1,2 \tag{3.40}$$

$$O_{1,i} = \mu_{B_{i-2}}(y) \text{ for } i = 3,4 \tag{3.41}$$

where x, y are the crisp inputs to node i, and A_i, B_i (small, large, etc.) are the linguistic labels characterized by appropriate membership functions μ_A, μ_B, respectively.

Layer 2: rule nodes. In the second layer, the AND operator is applied to obtain one output that represents the result of the antecedent for that rule, i.e., firing strength. Firing strength means the degrees to which the antecedent part of a fuzzy rule is satisfied and it shapes the output function for the rule. Hence the outputs $O_{2,i}$ of this layer are the products of the corresponding degrees from Layer 1.

$$O_{2,i} = w_i = \mu_{A_i}(x_1) \times \mu_{B_i}(x_2) \text{ for i=1,2} \tag{3.42}$$

Layer 3: average nodes. In the third layer, the main objective is to calculate the ratio of each ith rules firing strength to the sum of all rules firing strength. Consequently, wi is taken as the normalized firing strength.

$$O_{3,i} = \overline{w}_i = \frac{w_1}{w_1 + w_2} \text{ for i=1,2} \tag{3.43}$$

Layer 4: consequent nodes. The node function of the fourth layer computes the contribution of each ith rules toward the total output and the function defined as:

$$O_{4,i} = \overline{w}_i f_i = \overline{w}_i (p_i x_1 + q_i x_2 + r_i) \tag{3.44}$$

where wi is the ith nodes output from the previous layer. As for {pi, qi,ri}, they are the coefficients of this linear combination and are also the parameter set in the consequent part of the Sugeno fuzzy model.

Layer 5: output nodes. The single-node computes the overall output by summing all the incoming signals. Accordingly, the defuzzification process transforms each rule's fuzzy results into a crisp output in this layer:

$$O_{5,i} = \sum_i \overline{w}_i f_i = \frac{\sum_i w_i f_i}{\sum_i w_i} \quad (3.45)$$

This network is trained based on supervised learning. So our goal is to train adaptive networks to be able to approximate unknown functions given by training data and then find the precise value of the above parameters. The distinguishing characteristic of the approach is that ANFIS applies a hybrid-learning algorithm, the gradient descent method and the least-squares method, to update parameters. The gradient descent method is employed to tune premise non-linear parameters ({ai, bi, ci}), while the least-squares method is used to identify consequent linear parameters ({pi, qi,ri}).

The task of the learning procedure has two steps: In the first step, the least square method is to identify the consequent parameters, while the antecedent parameters (membership functions) are assumed to be fixed for the current cycle through the training set. Then, the error signals propagate backwards. Gradient descent method is used to update the premise parameters, through minimizing the overall quadratic cost function, while the consequent parameters remain fixed. The detailed algorithm and mathematical background of the hybrid-learning algorithm can be found in (Jang, 1993).

Demirkaya (2010) proposed the adaptive neuro-fuzzy inference system (ANFIS) in the construction of daily displacement forecasting system for the Schlegeis Dam. Data spanning from 1992 to 1988 of daily records of water level, concrete temperatures and the pendulum measurements were used for the analysis. Results showed that the ANFIS provided a high accuracy, reliability and prediction of displacements of the dam with correlation coefficients greater than 0.99.

Rankovic et al (2012) identified the non-linear structural behaviour of a dam using Adaptive network-based Fuzzy Inference System (ANFIS) models. These were used to predict radial displacements of Bocac Arch Dam covering over 11 years. The performance of these models was tested using correlation coefficients, mean absolute error and mean square error. The results showed that the developed models can provide accurate results as measured by the correlation coefficient r = 0.988.

HongZhong and XueHong (2012) formulated relationships between reservoir water level, temperature and crack open displacements of Chencun Concrete Dam in China using ANFIS models. These models were used to get rules of the adverse load combination that may affect the opening of cracks in dams. After analysis, results showed that the worst combination is the low water level and low air temperature. For dam safety, the authors suggested that medium water level and air temperature is required for no crack opening.

Bui et al (2016) proposed a hybrid model called swarm optimized neural fuzzy inference system in the prediction of horizontal displacements of Hoa Binh Hydropower Dam located in Vietnam. The length of the data used in the analysis was 11 years. The results obtained from the proposed methodology were compared with support vector regression, neural networks, Gaussian processes and random forests. It was concluded that the developed model can be used with high accuracy in displacement modelling.

3.2.5 Other methods

3.2.5.1 Blind Source Separation (BSS)

This is one of the methods that is used in data analysis and signal processing. This method recovers the underlying sources by exploiting the assumption of their mutual independence.

The BSS model is a linear simultaneous mixture given as: -

$$x = As + n \tag{3.46}$$

Where $x = [x_1, x_2, x_3 \ldots \ldots . x_m]^T$ is a vector containing measured scalar signals,

$s = [s_1, s_2, s_3 \dots \dots . s_n]^T$ is a vector containing sources $(m \geq n)$, A is an unknown mixing matrix of size mxn and n is the additive measurement noise. The main purpose of BSS is to find a demixing matrix A^{-1} such that $A^{-1}x$ recover the underlying sources s using the measured x. Figure 3.10 below shows a BSS process

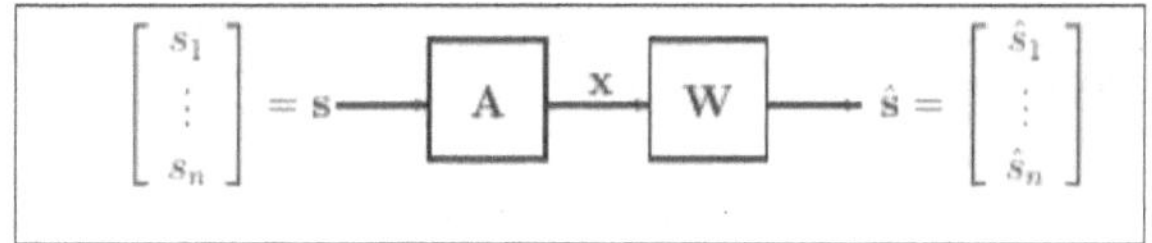

Figure 3.9. BSS model (Popescu 2011)

There exist two major BSS methods that are widely used namely: -

(i) higher-order statistics based on independent component analysis (Fantinato et al., 2019): this exploits the marginality non-Gaussianity and statistical independence of sources without considering the time structure of sources.

(ii) second-order statistics (SOS) based on second-order blind identification (SOBI) (Popescu 2011) which utilises SOS by considering multiple co-variance matrices of the responses, for the case when the sources are spatially uncorrelated but temporally correlated.

Some assumptions are made while doing a BSS analysis. These include: -

(i) The sources are mutually independent. The joint probability density of the sources should be factorizable into the product of their marginal densities.

(ii) Sources are to be non-Gaussian

(iii) The mixing matrix A to be full column rank

Popescu (2011) analysed a 1,200-days data from dam monitoring and surveillance of the Vidraru Dam in Romania days using blind source separation. This research was carried out to determine the influence of the external loads namely: air temperature and hydrostatic pressure on structural deformation and separate these external loads of the dam and the time effects without prior knowledge of the propagation environment. Second-order blind source separation (BSS) algorithm was used for the establishment of the independent sources and identification of the irreversible component in the structural response of the dam. Results from the analysis suggested that the blind source separation method can be effective in predicting dam behaviour.

3.2.6 Spectral based methods

3.2.6.1 Wavelet transforms

According to Staszewski (1997), the wavelet transform is a non-stationary signal by transforming its input time domain into a time-scale domain. Mathematically, wavelet transforms are inner products of the signal and a family of wavelets (Meyer 1993). Let ψ (t) be the analysing wavelet called the mother wavelet of the analysis. The corresponding family of wavelets consists of a series of son wavelets, which are generated by dilation and translation from the mother wavelet ψ (t) as shown in Equation 3.47 below:

$$\psi_{a,b}(t) = \frac{1}{\sqrt{a}}\psi\left(\frac{t-b}{a}\right), a > 0, b \in R \tag{3.47}$$

where a is the dilation or scale parameter defining the support width of the son wavelet and b is the translation parameter localizing the son wavelet function in the time domain. The function 1/a is used to establish energy preservation in the wavelet transform. The difference between these wavelets is due mainly to the different lengths of filters which define the wavelength and scaling functions. Wavelets must be oscillatory, must decay quickly to zero and must integrate to zero (Chui, 1992).

Let's decompose a signal x (t) into wavelet coefficients $W\psi$(a, b) using the basis of son wavelets ψa, $b(t)$. Under the hypothesis that x (t) satisfies the condition:

$$\int_{-\infty}^{+\infty} |x(t)|^2 dt < \infty \tag{3.48}$$

which implies that $x(t)$ decays to zero as the wavelet transform of $x(t)$ is expressed by the following inner product in the Hilbert space:

$$W_\psi(a,b) = \langle x(t), \psi_{a,b}(t)\rangle = \int_{-\infty}^{+\infty} x(t)\psi^*_{a,b}(t)dt \qquad (3.49)$$

where the asterisk stands for a complex conjugate. This indicates that the wavelet transform is a linear scalar product normalized by the factor 1/√a which measures the fluctuation of the signal $x(t)$ around point b at the scale a. The scaling operation executes stretching and compressing on the son wavelet, which can, in turn, be used to obtain the different frequency information of the signal to be analysed. The compressed version is used to meet the high-frequency needs, and the dilated version is used to meet the requirements of low-frequency. The translated version is used to get the time information about the signal to be analysed. As a result, a family of scaled and translated wavelets is generated and serves as the base for representing the signal to be analysed. This means the wavelet transform $W\psi$(a, b) can be considered as functions of translations *b* with each fixed scale *a*. It is also worth noting that the wavelet transform represents the convolution between the signal $x(t)$ and the wavelet function. This means that if the wavelet used is identical to the feature components (eigenfrequencies and damping coefficients) concealed in the analyzed signal, a wavelet may be used for feature discovery. For the function $\psi(t)$ to qualify as an analyzing wavelet, it must satisfy the admissibility condition (Chui, 1992):

$$0 < c_\psi = \int_{-\infty}^{+\infty} \frac{|\psi(\omega)|^2}{\omega} d\omega < \infty \qquad (3.50)$$

Where the Fourier transform is of ψ(ω). Then the wavelet transform can be inverted and the signal $x(t)$ recovered:

$$x(t) = \frac{1}{c_\psi} \int_{-\infty}^{+\infty} \int_{-\infty}^{+\infty} W_\psi(a,b)\psi_{a,b}(t) \frac{dadb}{a^2} \qquad (3.51)$$

Note that since tends to zero when tends to ±∞, the Fourier transform of the wavelet can be considered as a band-pass filter.

For practical purposes, the probability of time-frequency localization arises if the wavelet is a window function, which means that decays to zero as $t \rightarrow \pm\infty$:

$$\int_{-\infty}^{+\infty} |\psi(t)| dt < \infty \tag{3.52}$$

and the wavelet transform analyses a signal $x(t)$ only at windows defined by the wavelet function If one assumes a fast decay of the values are negligible outside a given time domain interval, the transform becomes local in the time domain, in this interval.

The frequency localization can be estimated when the wavelet transform is expressed in terms of the Fourier transform. Note $X(\omega)$ the Fourier transform of the signal and the Fourier transform of the son wavelet. Using Parseval's theorem (Meyer, 1993), we obtain

$$W_{\psi}(a, b) = \frac{\sqrt{a}}{2\pi} \int_{-\infty}^{+\infty} X(\omega) \psi^{*}(a\omega) e^{j\omega b} d\omega \tag{3.53}$$

and the frequency localization depends on the scale parameter a. Note that this operation is equivalent to a particular filter band analysis in which the relative frequency bandwidth is constant and related to the parameters a, b and to the frequency properties of the wavelet.

The local resolution of the wavelet transform in time and frequency is determined by the duration and bandwidth of the analysing functions given by $\Delta t = a\Delta t_{\varphi}$ and $\Delta f = \Delta f_{\varphi}/a$ where Δt_{φ} and $\Delta f_{\varphi}$are the duration and bandwidth of the wavelet function. The resolution of the analysis is therefore good for high dilation in the frequency domain and for low dilation in the time domain. For Morlet analysing wavelet function, the relationship between the dilatation parameter a and the frequency f at which the analysing wavelet function is focused is given by:

$$a = \frac{f_o f_s}{f_w} \tag{3.54}$$

where f_o is the frequency of the wavelet, f_s and f_w are the sampling frequencies of the signal and wavelet respectively.

Behr et al (1998) used spectral methods to study the relationship between air temperature and horizontal displacements of Pacoima Dam located in Los Angeles, California, USA. This study was started after the dam experienced an eastward displacement at the centre when air temperatures increased and westwards displacement when air temperatures are low. This led to the authors to understand the thermos-elastic behaviour of the dam accurately. Results showed that residuals from the analysis of deformations were related to reservoir water level changes and could be used for detection of any anomalies in the dam behaviour.

Su and Wu. (2007) developed a methodology of analysing dam monitoring data after a continuous crack was observed along the downstream wall of the upper gallery of the dam wall after ten years of operation. They proposed an algorithm for forecasting the nonlinear relation between loads and the behaviour of an arch-gravity dam using wavelet network. Vertical displacements, dam crack opening data, air temperature and water levels spanning for 12 years were used in the analysis. The wavelet network model was compared with the HST model and results showed that then the proposed method is feasible and effective in forecasting the dam behaviour.

Liu and Lian (2011) proposed a methodology based on wavelet transform to detect outliers in monitoring data collected from the Lijiaxia Concrete Dam located in China. The displacement time series data was decomposed into four levels using wavelet db1. Analysis of the data showed that wavelets can detect outliers in the time series which is a very important step in dam safety.

Mata and Santos (2016) developed statistical models coupled with wavelet transform in the analysis of displacements of Varosa Dam located in Portugal. Monitoring data ranging from November 2012 to April 2015 was subjected to the proposed methodology to understand the effect of daily variation effects of air temperature on the displacements of the dam. The authors indicated that wavelet transforms coupled with statistical models can extract information that cannot be obtained easily with traditional statistical models. It was also observed that anomalies can be detected early by understanding the daily effect of air temperatures on dam behaviour.

3.3 Review of data analysis techniques for dam dynamic properties

3.3.1 Introduction

Dams are subjected to vibrations from ground motion, water waves, and wind. In response to these forces, it was recommended by ICOLD that the performance of dams under vibrations should be evaluated (Wieland and Kirchen, 2012). These vibrations may be attributed to water waves, wind, and ground motions. These vibrations affect structural properties such as stiffness and damping of dams. The structural properties dams are related to are the so-called dynamic properties (natural frequencies, damping ratio and mode shapes). Dynamic properties of dams are used for the following purposes:-

(i) Finite element model calibration: For example, an Italian gravity dam (Sevier and De Falco (2019), Ridracoli Dam (Buffi et al., 2017), Roode Elsberg Dam (Makha 2012, Vezi, 2014). However, finite element modelling has several disadvantages such as the models may not represent the exact physical properties of the dam (Lotfollation-Yaghin and Heasri 2008, Yigit et al., 2016, Buffi et al., 2017). Therefore the real structural behaviour of these structures needs to be verified using field tests and continuous dynamic monitoring.

(ii) Structural health monitoring: tracking the dynamic properties through field tests and continuous dynamic monitoring such that if there is any change in the dynamic properties, it would indicate a change in the structural property of the dam.

3.3.2 History of dynamic monitoring of dams

Dynamic tests of dams are reported to have started around the late 1970s, for example, vibration tests were carried out on Emmoson Arch Dam in Switzerland (Deiunum et al., 1982). Also, several dams in the United Kingdom and Switzerland were subjected to vibration tests in early 1980 (Severn et al., 1981). The vibration tests performed on these dams were done using mechanical exciters such as servo-hydraulic shakers (Figure 3.10) and eccentric mass shakers (Figure 3.11). These shakers force the dam structure to vibrate such that responses from the dam can be measured. This type of vibration test is known as a forced vibration test (FVT).

Case studies among others where forced vibration tests were done include Wimble Ball Dam (Severn et al., 1980), Xiang Hong Dian (Clough et al., 1986) and Norsjo Dam (Cantieni, 2001), Shahid Rajaee Dam (Tarinejad et al., 2014), Daniel Johnson Dam (Gauron et al., 2017) and Baixo Sabor Dam (Gomes et al., 2018).

Figure 3.10: Servo hydraulic shaker (Cantieni, 2001)

Figure 3.11. Eccentric mass shaker (Gauron et al., 2017)

Dams are commonly situated in remote areas where accessibility may be a problem and transporting the mechanical exciters to site become very complicated and costly. In such instances, ambient vibration testing (AVT) becomes the only realistic way of carrying out dynamic tests on dams. AVT has advantages over FVT, i.e. it is cheaper and the structure is tested in its operating conditions. In ambient vibration testing, only structural responses are measured and it uses environmental forces (wind, ground movement) to excite the structure. Instead of using the force signal as a reference, the response signal measured at one or more points (often called reference points) is used when estimating the dynamic properties. These dynamic properties of a dam are extracted directed from accelerations recorded by sensors mounted on a structure using a technique known as operational modal analysis (OMA). Figure 3.12 shows the system scheme of OMA.

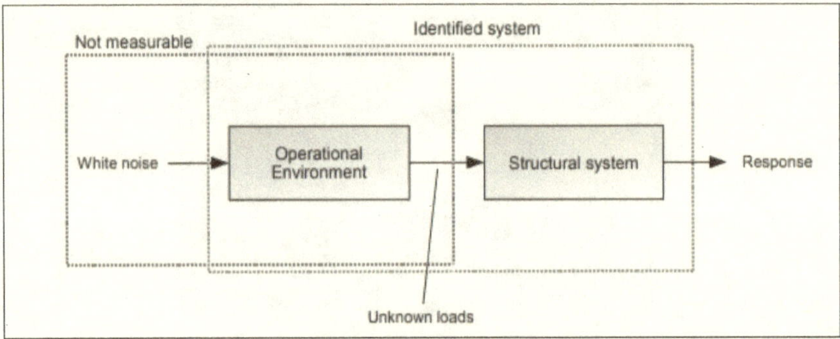

Figure 3.12. System scheme for OMA (Greiner 2009)

Several concrete dams around the world have been subjected to ambient vibration tests for example Contra Dam (Brownjohn et al., 1986), Hermitage Dam (Brownjohn, 1990), Fei Tsui Dam (Loh and Wu, 1996), Mauvoisin Dam (Darbre et al., 2002), Sahid-Rajaee and Saveh Dams (Mivehchi and Ahmadi, 2003), Gem Lake Dam (Ellis et al., 2010), Roode Elsberg and Kouga Dams (Moyo and Oosthuizen, 2010) and Berke Dam (Sevim et al., 2012). More information concerning the instruments and results obtained during the ambient vibration tests of Roode Elsberg Dam can be found in Bukenya et al, (2014).

3.3.3 Methods for analysing data from ambient vibration tests

Data collected from ambient vibration tests of dams is in the form of accelerations that need to be converted to the so-called dynamic properties (natural frequencies, damping ratio and mode shapes). There exist different methodologies that have been developed to estimate dynamic properties for the last two decades. These methodologies can either time domain, frequency domain or time-frequency domain (Zhang et al., 2005). This section discusses the different methods used in the estimation of the dynamic properties of dams.

3.3.3.1 Frequency domain-based

- **Peak picking**

The simplest method to estimate the modal parameters of a structure subjected to ambient loading is the basic frequency domain technique (Bendat and Piersol, 1993) or commonly known as peak picking method (PP). This technique is given the PP name because natural frequencies are obtained from the peaks of the power spectra density plots (Figure 3.13), obtained by converting the measured data to the frequency domain by Fast Fourier Transform (FFT). This method works on basic assumptions that around a resonance, only one mode is dominant and it has lower damping . When these assumptions are considered in the Equation (3.55) for the FRF and that the input spectrum is assumed to be constant, the output spectrum matrix is expressed as:

$$G_{yy}(j\omega)=\sum_{k=1}^{n}\sum_{s=1}^{n}\left[\frac{A_k}{j\omega-\lambda_k}+\frac{A_k^*}{j\omega-\lambda_k^*}\right]\times C\left[\frac{A_s}{j\omega-\lambda_s}+\frac{A_s^*}{j\omega-\lambda_s^*}\right]^H \tag{3.55}$$

If only one mode is dominant for example the rth mode then Equation (3.55) can be approximated as:

$$G_{yy}(j\omega)\approx\left[\frac{A_r}{j\omega-\lambda_r}+\frac{A_r^*}{j\omega-\lambda_r^*}\right] \tag{3.56}$$

where the residues are related to the mode shape. It means that, at resonance, each column of the spectrum matrix can be considered as an estimate of the corresponding mode shape, up to a scaling factor being the input unknown. To obtain such mode shape, however, the column of the spectrum matrix (and, therefore, the reference sensor) must be chosen so that it carries information about that mode: equivalently, the reference sensor cannot be a sensor placed at a node of the mode shape. Therefore, a good choice for the reference sensor allows the computation of only the spectra between all sensors and the reference instead of the full spectrum matrix.

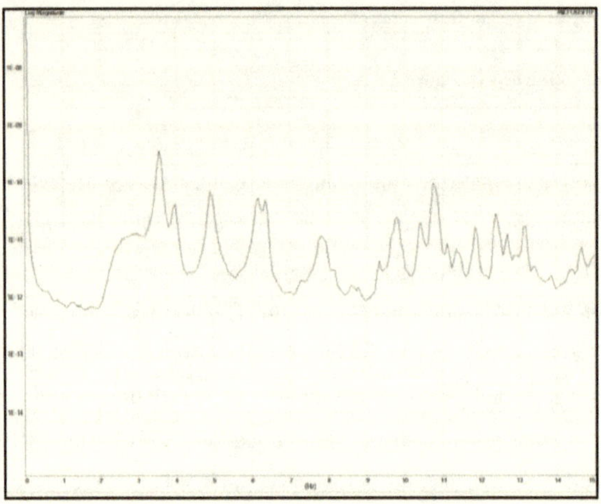

Figure 3.13: Power Spectral Density plot

Modal damping can be estimated using the half-power method (Choptra, 1995). The half-power method is not an accurate estimate for damping (Peeters, 2000).

The advantage of the PP technique is that it is simple and not demanding from a computational point of view. The main drawbacks of the method are (i) identifying operational deflection shapes instead of mode shapes and when closely spaced such an operational deflection shape will be the superposition of multiple modes (He and Fu, 2001) and (ii) the selection of the natural frequencies is a rather subjective task. The disadvantages of this method have been overcome by the development of frequency-domain decomposition techniques.

- **Frequency domain decomposition (FDD)**

The FDD approach is the spine of the frequency domain decomposition-type methods. This method uses the output PSD as the PP method but it carries out singular value decomposition (SVD) of the output PSD, estimated at discrete frequencies $\omega = \omega_i$ (Brincker et al., 2000) This decomposition is performed to identify a single degree of freedom models of the system (Batel, 2002)

The singular value decomposition of an m x n complex matrix A is the following factorization:

$$A = U\Sigma V^{H}$$

(3.57)

Where U and V are unitary and Σ is a diagonal matrix that contains the real singular values.

$$\Sigma = diag(s_1,....,s_r)$$
$$r = \min(m,n)$$

(3.58)

The superscript H on the matrix V denotes a Hermitian transformation (transpose and complex conjugate). In the case of real-valued matrices, the V matrix is only transposed. The s_i elements in the matrix S are called the singular values and their following singular vectors are contained in the matrices U and V.

This singular value decomposition is performed for each of the matrices at each frequency and for each measurement taken. The spectral density matrix is then approximated to the following expression after SVD decomposition (Brincker et al., 2000):

$$[G_{yy}(j\omega)] = [\Phi][\Sigma][\Phi]^{H}$$

(3.59)

with $$[\Phi]^{H}[\Phi] = [I]$$

$[\Sigma]$ being the singular value matrix and $s_1,\ldots, s_r$ the singular vectors unitary matrix:

$$[\Sigma] = diag(s_1,...s_r) = \begin{bmatrix} s_1 & 0 & 0 & . & . & 0 \\ 0 & s_2 & 0 & . & . & . \\ 0 & . & s_3 & . & . & . \\ . & . & . & . & . & 0 \\ . & . & . & & s_r & 0 \\ 0 & . & . & 0 & 0 & 0 \end{bmatrix}$$

(3.60a)

$$[\Phi] = [\{\varphi_1\}\{\varphi_2\}\{\varphi_3\}..\{\varphi_r\}]$$

(3.60b)

The number of non-zero elements in the diagonal of the singular matrix corresponds to the rank of each spectral density matrix. The singular vectors correspond to an estimation of the mode shapes and the corresponding singular values are the spectral densities of the SDOF system expressed in Equation (3.60b).

In the repeated mode case, the rank of the PSD matrix will be equal to the number of the multiplicity of the modes. Therefore, the SV function can be used as a modal indication function (MIF). Natural frequencies can be located by peaks of the SV plots. From the corresponding singular vectors, mode shapes can then be estimated. Since SVD can separate signal space from noise space, the modes can be indicated from SV plots, and closely spaced modes or even repeated modes can easily be detected (Zhang et al., 2005).

The FDD has the advantages of calculating closely spaced modes, being user-friendly but it cannot give modal damping ratios (Zhang et al., 2005). Therefore, the second generation of FDD-type method called Enhanced Frequency Domain Decomposition (EFDD) was proposed. This estimates not only modal frequencies and mode shapes but also damping ratios (Brinker et al., 2001).

Darbre et al. (2000) investigated the effect of different water levels on natural frequencies of the arch dam of Mauvoisin in Switzerland. Ambient vibration tests on the Mauvoisin Dam marked the start of long-term ambient vibration monitoring of dams. Between 1995 and 1996, seven ambient vibration tests at various water levels were carried out. An automated system was mounted on the dam and ambient vibrations were recorded twice daily for 6 months. Natural frequencies were obtained by peak picking on normalized power spectral densities of individual acceleration. Results indicated that the natural frequencies of the dam increased initially with rising water level and then decreased with further rise. This was contributed to the two competing features of increasing mass of water (reduction of natural frequencies) and dam stiffening due to closing of vertical construction joints.

Altunisik et al., (2015) investigated the effect of retrofitting on the dynamic characteristics of a laboratory arch dam model. Ambient vibration tests were carried out to illustrate changes in the dynamic characteristics of the arch dam model. Frequency domain decomposition was used in the extraction of natural frequencies from the ambient vibration measurements. Results showed that retrofitting led to the increase in natural frequencies indicating that repairing and strengthening of dams can restore dynamic characteristics.

3.3.3.2 Time domain-based techniques

- **Stochastic subspace identification**

Modal parameters from acceleration data collected from ambient vibration tests of Roode Elsberg dam were extracted using data-driven stochastic subspace identification method. Stochastic subspace identification is commonly used when attempting output-only modal analysis on structures in operation; it is thought to be one of the most powerful identification techniques currently available for modal parameters (Peeters and De Roeck, 2001, Brinker and Andersen, 2006). Estimates of modal parameters are obtained via the identification of a discrete state-space model where the (ambient) inputs to the system are assumed to be white noise.

The SSI-data driven approach involves the identification of the state-space model which is achieved by the construction of a block Hankel matrix of measurement data and employment of numerical techniques such as singular value decomposition and QR factorisation. For more details readers should consult Van Overschee, P.; De Moor, B. (1996), where examples of code for the implementation of SSI routines are available.

The system can be considered in the classical formulation as a multi-degree of freedom structural system (Andersen, 2010)

$$M\ddot{y}(t) + C\dot{y}(t) + Ky(t) = F(t) \tag{3.61}$$

Where M is the mass matrix, C is a damping matrix, K is the stiffness matrix and F(t) is the force vector.

The stochastic response from a system can be expressed as a function of time

$$y(t) = \begin{Bmatrix} y_1(t) \\ y_2(t) \\ . \\ . \\ y_m(t) \end{Bmatrix} \tag{3.62}$$

Introduction of the state space formulation takes the classical continuous-time formulation to the discrete-time domain as follows:

$$x(t) = \begin{Bmatrix} y(t) \\ \dot{y}(t) \end{Bmatrix} \tag{3.63}$$

The second-order system equation is given by Equation (3.63) and can be simplified to a first-order equation by introducing the state-space formulation:

$$\dot{x}(t) = A_c x(t) + B_c f(t)$$
$$y(t) = Cx(t) \tag{3.64}$$

Where the system matrix A_c in continuous time and load matrix B_c is given by

$$A_c = \begin{bmatrix} 0 & I \\ -M^{-1}K & -M^{-1}C \end{bmatrix}$$

$$B_c = \begin{bmatrix} 0 \\ M^{-1} \end{bmatrix} \tag{3.65}$$

Kailath (1980) showed that this formulation has an advantage in that the general solution is directly available for example

$$x(t) = \exp(A_c t)x(0) + \int_0^t \exp(A_c(t-\tau))Bf(\tau)d\tau \tag{3.66}$$

Where the first term is the solution to the homogeneous equation and the last term is the solution. To make this solution to discrete-time, all variables are sampled like $y_k = y(k\Delta t)$ and thus the solution to the homogenous equation becomes (Brincker and Andersen, 2006)

$$\begin{aligned} x_k &= \exp(A_c k\Delta t)x_0 = A_d^k x_0 \\ A_d &= \exp(A_c \Delta t) \\ y_k &= C A_d^k x_0 \end{aligned} \tag{3.67}$$

This construction is simply defined by its power series and is calculated in practice by performing an eigenvalue decomposition of the involved matrix and then taking the exponential function of the eigenvalues. Note that the system matrix in continuous time and discrete in time is not the same.

The Block Hankel matrix

In discrete-time, the system response is normally represented by the data matrix

$$Y = \begin{bmatrix} y_1 & y_2 & \dots & y_N \end{bmatrix} \tag{3.68}$$

Where N is the number of data points.

Consider a more simple case where we perform the product between two matrices that are modifications of the data matrix given by Equation (3.75). Let $Y_{(1:N-k)}$ be the data matrix where we have removed the last k data points, and similarly, let $Y_{(k:N)}$ be the data matrix where we have removed the first data points, then

$$R_k = \frac{1}{N-k} Y_{(1:N-k)} Y_{(k:N)}^T \tag{3.69}$$

This is an unbiased estimate of the correlation matrix at time lag k. This follows directly from the definition of the correlation estimate, see for instance Bendat et al., (1986) The Block Hankel matrix defined in SSI is simply a gathering of a family of matrices that are created by shifting the data matrix

$$Y_h = \begin{bmatrix} Y_{(1:N-2s)} \\ Y_{(2:N-2s+1)} \\ \vdots \\ Y_{(2s:N)} \end{bmatrix} = \begin{bmatrix} Y_{hp} \\ Y_{hf} \end{bmatrix} \tag{3.70}$$

The upper half part of this matrix is called "the past" and denoted Y_{hp} and the lower half part of the matrix is called "the future" and is denoted Y_{hf}. The total data shift is $2s$ and is denoted "the number of block rows" (of the upper or lower part of the Block Hankel matrix).

The Projection

OMA deals with stochastic responses where projection can be defined as a conditional mean (Overschee and De Moor, 1996). It is useful in combining the future into the past defined by the matrix

$$O = E(Y_{hf} | Y_{hp}) \tag{3.71}$$

Melsa and Sage (1973) suggested that such a conditional mean can be described by its covariance for Gaussian processes. Since the shifted data matrices also define covariances, it is not so strange that the projection can be calculated directly as also defined by van Overschee and De Moor (1996).

$$O = Y_{hf} Y_{hp}^T (Y_{hp} Y_{hp}^T)^{-1} Y_{hp} \quad (3.72)$$

The last matrix in this product defines the conditions, whereas the first four matrices in the product introduce the covariances between channels at different time lags. A conditional mean as given by Equation (3.78) simply consists of free decays of the system given by different initial conditions specified by Y_{hp} (Brincker and Andersen, 2006). The matrix is *sM x sM* and any column in the matrix O is a stacked free decay of the system to a (so far unknown) set of initial conditions. Using Equation (3.75) any column in O can be expressed by

$$O_{col} = \Gamma_s x_0$$
$$\Gamma_s = \begin{bmatrix} C \\ CA_d \\ CA_d^2 \\ \vdots \\ CA_d^{s-1} \end{bmatrix} \quad (3.73)$$

Now, if we knew the so-called observability Γ_s matrix, then we could simply find the initial conditions directly from the Equation (3.72) (it is a useful exercise to simulate a system response from the known system matrices, use the SSI standard procedure to find the matrix **O** and then try to estimate the initial conditions directly from Equation (3.72)).

The Kalman States

The so-called Kalman states are simply the initial conditions for all the columns in the matrix O, thus

$$\mathbf{O} = \Gamma_s X_0 \quad (3.74)$$

Where the matrix X_0 contains the so defined Kalman states at time lag zero. Again, if we knew the matrix Γ_s, then we could simply find all the Kalman states directly from Equation (3.79), however, since we do not know the matrix Γ_s , we cannot do so, and thus we should estimate the states differently. The trick is to use the SVD on the O matrix

$$O = USV^T \tag{3.75}$$

And then define the estimate of the matrix Γ_s and the Kalman state matrix X_0 states by

$$\begin{aligned}\hat{\Gamma} &= US^{1/2}\\ \hat{\mathbf{X}}_0 &= S^{1/2}V^T\end{aligned} \tag{3.76}$$

The so defined procedure for estimating the matrices $\hat{\Gamma}$ and $\hat{\mathbf{X}}_0$ is not unique. A certain arbitrary similarity transformation can be shown to influence the individual matrices, but can also be shown not to influence the estimation of the system matrices.

The Kalman state matrix $\hat{\mathbf{X}}_0$ is the Kalman state matrix for time lag zero. If we remove one block row of O from the top, and then one block row of $\hat{\Gamma}$ from the bottom, then similarly we can estimate the Kalman state matrix $\hat{\mathbf{X}}_1$ at time lag one. Thus by subsequently removing block rows from O, all the Kalman states can be defined. Using the Kalman states a more general formulation for estimating also the noise part of the stochastic response modelling can be established. However, in this book, the focus is on explaining how the system matrices can be found, and in this context, there is no further need for Kalman states.

The system matrix A_d can be found from the estimate of the matrix Γ by removing one block from the top and one block from the bottom yielding

$$\hat{\Gamma}_{(2:s)}\hat{A}_d = \hat{\Gamma}_{(1:s-1)} \tag{3.77}$$

And thus, the system matrix $\hat{A}_d$ can be found by regression. The observation matrix C can be found simply by taking the first block of the observability matrix

$$\hat{C} = \hat{\Gamma}_{(1:1)} \tag{3.78}$$

Determination of modal parameters

The first step of finding the modal parameters is to perform an eigenvalue decomposition of the system matrix $\hat{A}_d$

$$\hat{A}_d = \Psi[\mu_i]\Psi^{-1} \tag{3.79}$$

The continuous-time poles λ_i are found from the discrete-time poles μ_i by

$$\mu_i = \exp(\lambda_i) \tag{3.80}$$

where

$$\lambda_i = \frac{In(\mu_i)}{\Delta T}$$
$$\omega_i = |\lambda_i|$$
$$\varsigma_i = \frac{\mathrm{Re}(\lambda_i)}{|\lambda_i|} \tag{3.81}$$

The mode shape matrix is found from

$$\Phi = C\Psi \tag{3.82}$$

Andersen and Brincker (2010) described three SSI algorithms namely Unweighted Principal Component (UPC), Principal Component (PC) and Canonical Variate Analysis (CVA). These SSI algorithms are described with the use of two weight matrices W_1 and W_2 applied to O_i. Still, O_i is common to all three algorithms (Andersen and Brincker, 2010). Thus, it is also called the common SSI input matrix.

$$W_1 O_i W_2 = W_1 \Gamma_i \hat{X}_i W_2$$
$$= [U_1 \quad U_2]\begin{bmatrix} S_1 & 0 \\ 0 & 0 \end{bmatrix}\begin{bmatrix} v_1^T \\ v_2^T \end{bmatrix}$$

$$=U_1 S_1 V_1^T \tag{3.83}$$

The above-mentioned methodologies have been used to estimate the dynamic properties of dams around the world. Bukenya et al., (2014) presented several examples where different methods were used to extract natural frequencies from data collected from ambient vibration surveys.

Okuma et al. (2008) collected basic data for developing structural damage detection based on long term ambient vibration testing of the Hitotsuse Arch Dam in Japan. Long term vibration monitoring was conducted to determine the changes in the natural frequencies of the whole dam due to macroscopic damage caused by ageing effects. The measuring system was designed to record continuously at a sampling rate of 200 Hz and store data measured every 30 minutes. Natural frequencies were estimated from the cross-spectrum of autoregressive moving average models. Results indicated that the identified natural frequencies were in good agreement with the earthquake observation records, and the identified natural frequencies of three modes strongly correlated with the water level of the dam.

Oliviera et al., (2010) described results from a continuous dynamic monitoring system installed at the Cabril dam for a year. The objective of this study was to study the correlation between changes in the modal parameters and structural changes due to deterioration processes, especially between the development of cracking and changes in modal shapes and to study the influence of the reservoir on the structural dynamic behaviour of the dam-foundation-reservoir system. Modal parameters were estimated using SSI-data driven method. Results showed that it was possible to automatically identify the modal parameters of the dam at an hourly basis. This, in turn, helps in detecting any small changes in the modal parameters that are influenced by changes in water level.

Pereira et al (2017) presented results from dynamic monitoring data collected from Baixo Sabor dam located in Portugal for the first six months of operation. The purpose of this study was to investigate the effect of water level on the modal parameters of the dam. This, in turn, will help in the development of statistical models. COV-SSI was used in the estimation of modal parameters. Initial results showed that water level influences the natural frequencies of the dam.

3.4 Chapter summary

Monitoring the static and dynamic behaviour of large dams is a subject of great importance because of the effect such structures have on the landscape where they have been constructed. Health monitoring of dams plays a significant role in ensuring their structural integrity and maintaining their longevity. Dam health monitoring can be divided into two namely: static and dynamic monitoring.

This chapter has presented a literature review of the techniques used in the analysis data collected from static and dynamic monitoring of dams. Case studies where each technique was applied were presented. Static monitoring of dams includes measurement of variables such as ambient temperatures, reservoir level, opening and closing of joints, the opening of crack, displacements and strains which are measured accurately by instruments mounted on dam walls. In the analysis of dam monitoring results, mainly statistical models (HST model and multivariate statistical methods) were used, although other methods were implemented as structural identification techniques, short-term Fourier transform and blind source separation. The main reason why statistical methods have been widely used is that they are simple in the formulation and provide any correlation between governing and dependent parameters. In all the examples given, a lot of emphasis was put on the effect of reservoir water level, the temperature on deformations and most of them have been a success. It can also be seen that long-term data records are required for meaningful data analysis in dam surveillance. This can be seen that in the examples provided, most data exceeds 10 years, although there is one given in the analysis for less than 4 years. The review shows that static monitoring of dams has an outstanding track record and is very useful in assessing the structural health of the dam.

There has been dynamic monitoring of dams since the 1980s, with forced vibration tests being the first to be carried out on dams. Short term measurements of dynamic properties were done to mainly establish the relationship between reservoir water level and natural frequencies, then validate and calibrate analytical models. From the year 1990, ambient vibration testing became a popular approach due to developments in data acquisition systems technology and research in signal processing techniques. In the early 1990s, only once off field tests based on ambient vibration were carried out for several purposes such as; (1) investigation of the feasibility and suitability of AVT of dams (2) investigating the safety and stability of dams (3) determination of dynamic parameters for calibration and validation of finite element models of dams.

However, repeated measurements have been carried out on various dams since 2000 for; (1) investigation of the effect of reservoir levels on natural frequencies (2) development of a continuous dynamic monitoring system and (3) development of a long-term AVT-based damage detection system. This shows that ambient vibrations tests can now be used for long term safety monitoring of dams. Frequency domain methods (peak picking and frequency domain decomposition) is used for most of the case studies presented to estimate the dynamic properties from vibration measurements. This is because compared to time domain-based methods these methods are simple and easy to use.

Table 3.1 and 3.2 summarizes the review on structural health monitoring of dams including year, name of the dam, country, and method of analysis and output for dynamic and static monitoring respectively. Advantages and disadvantages of common data-driven models are shown in Table 3.3 below.

Table 3.1: Review summary of case studies and methods of analysis (Dynamic monitoring)

Name of the dam	**Type**	**Country**	**Method of analysis**	**Output**
Cabril	Arch	Portugal	FDD, SSI	Natural frequency
Hitsosuse	Arch	Japan	Peak picking	Natural frequency
Baixo	Arch	Portugal	SSI	Natural frequency

Table 3.2: Review summary of case studies and methods of analysis (static monitoring)

Name of the dam	**Type**	**Country**	**Method of analysis**	**Output**
Ferden	Arch	Switzerland	Probalistic model	Deformation
Roggiasca	Arch	Switzerland	Probabilistic model	Deformation
Idduki	Arch	India	HST	Strains
Schlegsis	Arch	Austria	NARX	Deformation
Landon	Gravity	Greece	DFT and LNP	Deformation

Daniel Johnson dam	Arch+buttress	Canada	HST	Deformation
Ancipa	Buttress	Italy		Deformation
?	Arch	China	RR, PCR and Step-wise regression	Deformation
Zillergrundl	Arch	Austria		Deformation
	Arch	China	BP-ANN	Deformation
Cheucen				
Alto Rabago	Arch	Portugal	MLR and NN	Deformation
Fei-Tsui	Arch	Taiwan	SSA-AR, NPCA and AANN	Deformation
Vidraru	Arch-buttress	Romania	BSS	Deformation
Cabril	Arch	Portugal	Regression	inclinations
Schlegsis	Arch	Austria	MLR and MLP	Deformation
Bocac	arch	Bosnia	ANFIS	Deformation
Tucurui dam	Gravity	Brazil	MLR and Control charts	Deformation
Wanfu	Arch	China	MLR and ECM	Deformation
Mianhuastan	Gravity	China	L-SVM	Deformation
Alto Lindoso	Arch	Portugal	SFTT	Deformation
Fei-Tsui	Arch	Taiwan	NARX	Deformation
Huaguantan			HSS	Deformation
	Gravity		SVM	Deformation
Djerdap		Serbia	SVR and NARX	Deformation
Panta La Baells	Arch	Spain	RF,BRT,NN,MARS,SVM	Deformation
Itaipu	Gravity	Brazil	PLS	Deformation
Gezhouba		China	PLSR	Deformation

Despite the rich literature that exists in the methodologies used in the analysis of dam monitoring data, there are still some issues that need to be addressed.

The literature reviewed suggested that there are few case studies in which various models were explicitly compared when predicting deformations on a common dataset. However, the relative merits of those models and their performance in the prediction of deformation have not been evaluated explicitly in a sufficiently objective manner.

As seen in Tables 3.1 and 3.2, only deformation monitoring utilises data-driven models in the prediction of dam responses. Therefore, there is a need to develop methodologies that predict natural frequencies of dams from ambient monitoring data using data-driven models.

There are no studies in the literature that integrated both deformation and ambient vibration monitoring in a single study. Therefore, there is a need to integrate ambient vibration monitoring data and deformation monitoring data. This will provide a complete assessment of structural performance of dams.

In conclusion, to address the above-mentioned issues, this study will involve the prediction of natural frequencies and deformations based on monitoring data. To achieve this, a methodology that gives better results than the methods summarised in Table 3.3 and can adapt to other variables other than deformations was proposed. To this effect, a machine learning-based methodology known as Gaussian Process Regression is used in the prediction of both deformations and natural frequencies. Since dam monitoring aims to detect any abnormal dam behaviour, there is a need to utilise methods that are robust in detecting any anomalies in monitoring data. Therefore, a methodology based on the least trimmed squares will be used. In the next chapter (Chapter 4), the methodology that was used in the analysis of monitoring data is described.

Table 3.3. Advantages and disadvantages of commonly used data-driven models in dam monitoring

Model	Advantages	Limitations
HST	• Easy to implement	• Effects due to thermal load are estimated without using the evolution of air temperature
HTT	• More accurate thermal effect due to the use of temperature data collected from embedded thermometers in the dam wall	• Criteria for choosing what thermometers are used in the model is unclear
HSS	• Accurate extraction of the time effect deformation	• Time-consuming
PLS	• Produces accurate results even when independent variables are more than dependent variables • Handles data with collinearity when independent variables are more than dependent variables	• Sensitive to the relative scaling of dependent variables
ICR	• Computationally efficient	• Source of ambiguity
AR	• Use of historical data to predict future trends	• Does not perform well with data that has anomalies
NARX	• Can deal with non-linear relationships between independent and dependent variables	• Inaccurate results when data has noise
ANFIS	• Deals with both numerical and categorical variables • Easily adapts to the nature of data • Can deal with non-linear relationships between independent and dependent variables	• Computationally expensive
MARS	• Deals with multivariate data • Can handle data with missing values	• Inaccurate results if governing variables are correlated • Functions need to be defined beforehand

ANN	• Deals with missing data • Deals with complex nonlinear relationships between dependent and independent variables	• Requires experienced users • Prone to overfitting • Computationally expensive
RF	• Deals with both numerical and categorical variables • Can handle data with missing values • Robust to outliers • Not affected by noisy data	• Complex • Requires a long training period
SVR	• Deals with nonlinear relationships • Risk of overfitting is less • Relatively memory efficient	• Choice of kernels is not direct • Inaccurate results if data has a lot of noise • Not good at modelling large data sets
BRT	• Simple to use • Requires less effort in data pre-processing • Handles data with missing values	• Sensitive to outliers • Computationally expensive

4 Methodology

4.1 Introduction

Instrumentation and monitoring aim to preserve and enhance the safety of the dam by providing information on 1) determining whether a dam is performing as expected and 2) ascertain if any dangers in case there is a change in the behaviour of the dam. There is a lot of data that is collected from dam monitoring and it is important to analyse this data to ensure that the principle objectives of monitoring are achieved. In Chapter 3, dam monitoring and data analysis focus mostly on deformations and no attention is given to vibration-based monitoring using ambient vibrations. Therefore, it is important to develop procedures that predict not only deformations but also natural frequencies of dams.

Therefore, this chapter discusses the methodologies namely (Gaussian process regression and least trimmed squares) that will be used to predict deformation and natural frequencies of Roode Elsberg Dam. Lastly, procedures used in comparing the performance of the proposed models and commonly used models in the analysis monitoring data are presented herein.

Figure 4.1 shows a flow chart of the methodology that is utilised in the analysis of the data. Four steps were followed namely:

(i) Creation of a data matrix that consisted of loads (temperature and water level) and the dam responses (deformations and natural frequencies). All measurements had the same measurement frequency i.e., the daily average of the observations. The gaps that were in the readings were filled in this step using forecasting of time series.

(ii) Canonical correlation analysis of the matrix created in step 1 to identify the associations between the loads and responses.

(iii) Statistical modelling of the dam responses using the methods shown in Figure 4.1 below. In this step, the performance of the chosen models is compared.

(iv) Residual analysis.

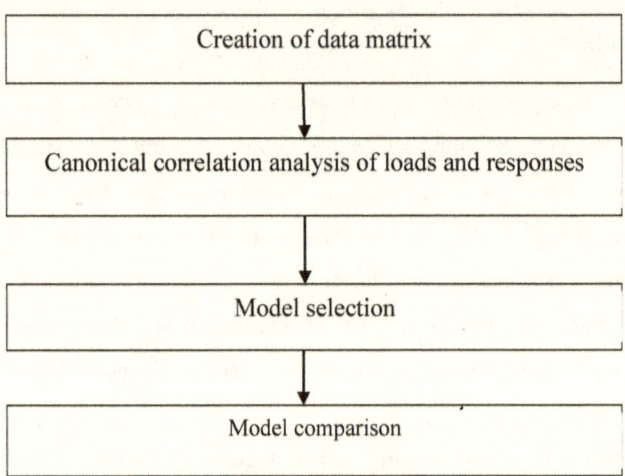

Figure 4.1. Flow chart of data analysis

4.1.1 Canonical correlation analysis

Canonical correlation analysis is a multivariate technique that identifies the independency between independent (X) and dependent variables (Y). This technique finds the correlation between a linear combination of the independent variables X and a linear combination of dependent variables Y to find the largest correlation between the two variables (Johnson and Wichern, 2007).

Let U and V be linear combinations of the vectors in X and Y respectively, then the task of canonical analysis is to obtain vectors of the coefficients a and b that maximise the correlation between U and V (Equation 4.2). In this case, U and V are known as canonical variables (Johnson and Wichern, 2007) as given in Equation 4.1 below

$$\begin{aligned} U &= a'X \\ V &= b'Y \end{aligned} \tag{4.1}$$

$$Corr(U,V) = \sqrt{\lambda} = \frac{a'\Sigma_{XY} b}{\sqrt{a'\Sigma_X a}\sqrt{b'\Sigma_Y b}} \tag{4.2}$$

The solutions of a and b form a system of equations shown in Equation (4.3) below.

$$
\begin{aligned}
\left(\Sigma_{XY}\Sigma_{Y}^{-1}\Sigma_{YX} - \lambda\Sigma_{X}\right)a = 0 \\
\left(\Sigma_{YX}\Sigma_{X}^{-1}\Sigma_{XY} - \lambda\Sigma_{X}\right)b = 0
\end{aligned}
\qquad (4.3)
$$

Where λ is the largest eigenvalue of the matrix.

The major aim of canonical correlation analysis is to obtain a few pairs of canonical variables that can explain much of the relationship between X and Y.

4.1.2 Model selection

In Chapter 3, it was shown that there are so many models that have been used in dam monitoring with most tested on understanding the effect of thermal and hydrostatic loads on dam deformation. However, there is a need to test the same models and how they are appropriate in understanding both the static and dynamic behaviour of dams. As seen in Chapter 3, several data-driven models have been used in the modelling of data collected from different dams. The choice of models to be used in this study becomes critical. In this regard, a machine learning algorithm called Gaussian process regression was chosen as this model had not been used before to the best of my knowledge in the prediction of dam responses especially natural frequencies. The performance of the selected GPR model was compared with other machine learning algorithms, that is to say, support vector regression (SVR), artificial neural network (ANN) and boosted regression trees (BRT). Also, GPR models were compared with other regression models namely, multivariate adaptive regression splines (MARS) and multiple linear regression (MLR). The above-mentioned models were compared with the widely used HST.

4.1.2.1 Variable selection

In the selection of the variables, engineering judgement is needed. According to Chapter 5, it can be seen that both deformation and natural frequencies are strongly dependent on the water level. Natural frequencies are strongly dependant on water level while deformations are strongly affected by temperature variations. The following variables were chosen.

- Variables measured on site
  (i) Daily water level
  (ii) Air temperature
  (iii)Water temperatures

- Derived variables
  (i) Moving average of air temperature over 30 days.

4.2 Gaussian Process Regression

Machine learning algorithms have been used in predicting the behaviour of dams (Salazar et al., 2015). This is because these algorithms perform better in non-linear relationships between dam responses and operational factors. Bukenya et al (2017) applied models to derive relationships between natural frequencies and operational factors (water level and temperature) from monitoring data collected from Roode Elsberg Dam. However, the challenge is the non-linear relationship between the responses and loads. The challenge is to develop models that can forecast dam natural frequencies accurately. This study proposes Gaussian Process regression models in an attempt to develop accurate models.

Gaussian Processes combine the Bayesian learning with kernel functions providing a principled and probabilistic approach for regression (Rasmussen and William, 2006). GP defines a prior probability distribution over latent functions directly specified by its mean function $m(\boldsymbol{x})$ and covariance (kernel) function, $k(\boldsymbol{x}, \boldsymbol{x}')$ (Rasmussen and William, 2006):-

$$f(x) \sim GP(m(\boldsymbol{x}), k(\boldsymbol{x}, \boldsymbol{x}')) \tag{4.5}$$

Where the mean function is assumed to be zero and it encodes tendency of the function (Zheng et al., 2016).

4.2.1 Fundamentals of Gaussian process regression

The relationship between input variables, x and output variables, y can be represented by a regression model represented as:-

$$y_i = f(x_i) + \varepsilon \tag{4.6}$$

Where f is the regression function ε is the error due to modelling assumed to be independent and ought to have a Gaussian distribution with zero mean and variance, σ_n^2 , that is:-

$$\varepsilon \sim N(0, \sigma_n^2) \tag{4.7}$$

From Equation (4.7), the likelihood is given by (Zhang et al., 2016):-

$$p(y|\boldsymbol{f},\boldsymbol{X}) = N(\boldsymbol{f},\sigma_n^2\boldsymbol{I}) \tag{4.8}$$

where I is the identity matrix, $\boldsymbol{f} = [f(\boldsymbol{x}_1), f(\boldsymbol{x}_2), \ldots\ldots f(\boldsymbol{x}_n)]^T$ which is assumed to be a Gaussian process (Teimouri et al., 2016):-

$$p(\boldsymbol{f}|\boldsymbol{X}) = N(0,\boldsymbol{K}) \tag{4.9}$$

Where K is the covariance (kernel) function whose element $K_{ij} = k(x_i, x_j)$ representing the covariance between $f(x_i)$ and $f(x_j)$ values.

The marginal distribution of y follows a Gaussian distribution and is given by (Zhang et al., 2016):-

$$p(\boldsymbol{y}) = \int p(y|\boldsymbol{f})p(\boldsymbol{f})d\boldsymbol{f} = N(\boldsymbol{0},\boldsymbol{K}+\sigma_n^2\mathbf{I}) \tag{4.10}$$

To forecast a new variable y* using a new x* value, the joint distribution of y and y is expressed as (Kang et al., 2017):-

$$\begin{bmatrix} y \\ y_* \end{bmatrix} \sim N\left(0, \begin{bmatrix} K(X,X)+\sigma_n^2\mathbf{I} & K(X,X_*) \\ K(X_*,X) & K(X_*,X_*) \end{bmatrix}\right) \tag{4.11}$$

According to Zhang et al., (2016), rules for conditions Gaussians, the predictive distribution $p(y|\boldsymbol{y})$ is a Gaussian distribution with mean and covariance expressed by

$$m(\boldsymbol{x}_*) = K(\boldsymbol{X}_*,\boldsymbol{X})[K(\boldsymbol{X},\boldsymbol{X})+\sigma_n^2\mathbf{I}]^{-1}\boldsymbol{y} \tag{4.12}$$

$$\sigma^2(\boldsymbol{x}_*) = K(X_*,X_*)[K(X,X)+\sigma_n^2\mathbf{I}]^{-1}K(X,X_*) \tag{4.13}$$

4.2.2 Covariance (kernel) functions

The results obtained from Equation (4.12) is affected by the covariance (kernel) function. This means that the choice of a kernel function is very important in obtaining accurate results in Gaussian process regression. There are different kernel functions used in Gaussian process regression (Rasmussen and Williams, 2006). Below is the description of the kernel functions commonly used:

(1) Squared Exponential

$$k(x_i, x_j|\theta) = \sigma_f^2 exp\left[-\frac{1}{2}\frac{(x_i-x_j)^T(x_i-x_j)}{\sigma_l^2}\right] \tag{4.13}$$

(2) Exponential

$$k(x_i, x_j|\theta) = \sigma_f^2 exp\left[-\frac{r}{\sigma_l}\right] \tag{4.14}$$

(3) Matern 3/2

$$k(x_i, x_j|\theta) = \sigma_f^2 \left(1 + \frac{\sqrt{3}r}{\sigma_l}\right) exp\left[-\frac{\sqrt{3}r}{\sigma_l}\right] \tag{4.15}$$

(4) Matern 5/2

$$k(x_i, x_j|\theta) = \sigma_f^2 \left(1 + \frac{\sqrt{5}r}{\sigma_l} + \frac{5r^2}{3\sigma_l^2}\right) exp\left[-\frac{\sqrt{5}r}{\sigma_l}\right] \tag{4.16}$$

(5) Rational quadratic

$$k(x_i, x_j|\theta) = \sigma_f^2 \left(1 + \frac{r^2}{2\alpha\sigma_l^2}\right)^{-\alpha} \tag{4.17}$$

Where $r = \sqrt{(x_i - x_j)^T(x_i - x_j)}$ is the Euclidean distance between x_i and x_j. σ_l and σ_f are the characteristic length scale and signal standard deviation respectively, α is the index. According to Rasmussen and Williams (2006), a gradient-based algorithm is used to determine the hyperparameters of covariance function $\theta(\sigma_l, \sigma_f)$.

4.3 Least trimmed squares based regression

In the analysis of monitoring data from dams, the developed models commonly used such as the HST are based on the least-squares method. However, in the presence of outliers, least square-based methods tend to suffer and give erroneous results. Least trimmed square (LTS) based regression, on the other hand, helps in dealing with data with outliers. The LTS method is a robust regression method which was proposed by Rousseeuw (1984). It minimizes the sum of squares of residuals and eliminates outliers if the data is big such that the predicted values are close to the measured values. The algorithm described herein was proposed by Rousseeuw (1984) known as the Fast least trimmed squares (Fast-LTS).

Consider a multiple linear regression

$$y = \beta X + \varepsilon \tag{4.18}$$

Where β are the regression coefficients, X is a nxmmatrix containing dependent variables and ε is the error. If the residual sum of squares resulting from regression is given by:

$$r = (y - X\beta) \tag{4.19}$$

The least trimmed squares regression is given as (Gilon and Padberg, 2002).

$$\min \sum_{i=1}^{h} (r^2)_{i:n} \tag{4.20}$$

Where h represents the number of points that are not trimmed from the data set and depends on a certain trimming factor α

Where the squares residual are ordered in an ascending order

$$(r^{2)})_{(1)} \leq .... \leq (r^{2)})_{(i)} \leq ...... \leq (r^2)_{(n)} \tag{4.21}$$

Then the objective of the Fast-LTS algorithm is to minimise the function resulting in the so-called LTS estimator (Hubert et al., 2005):-

The following steps are then followed:

1. Create an initial number of subsets h given by $h = (n + m + 1)/2$ where n is the number of observations and m is the number of independent variables.
2. Draw h-subsets H_1 and obtain h-subset H_2 by taking the h observations with smallest absolute residuals.

3. Apply least squares to H_2 to yield a new fit which will have a lower function (equation….)

The Fast-LTS algorithm has the following advantages (Rouswseuw P and Leory A 1987):

- The algorithm can deal with large data sets
- It is computationally inexpensive as compared to other methods such as the least median squares
- It has a smooth function, less sensitive in the presence of local effects.

4.4 Predicting model accuracy

Statistical models are constructed using a given set of data then these developed models are applied to a new set of data that were not used in the construction of the models. It becomes therefore imperative to measure the accuracy of the developed models on both past data and new data. The procedure involves dividing data into two sets namely: - a training set and testing test (Figure 6.3) although sometimes, data is further divided into a third set known as a validation set.

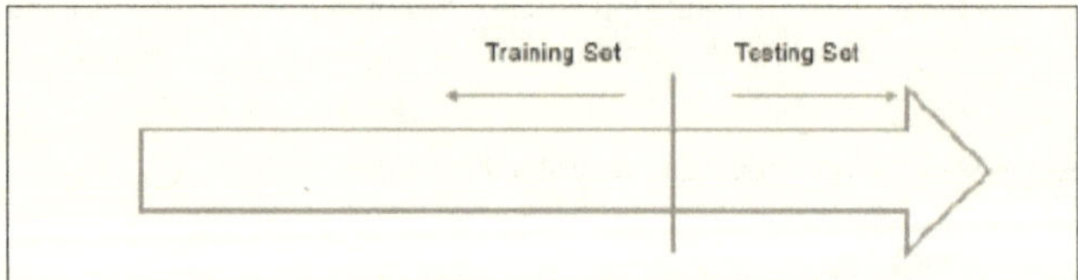

Figure 4.2. The sampling method used in the analysis

In dam monitoring, once regression models have been developed, there is a need to determine the prediction error of the estimated model. The reasons for calculating the prediction error include (Salaza, 2005):- (i) it provides an insight into its accuracy (ii) allows comparison of different prediction models and (iii) it defines thresholds. The following criteria were used in the measurement of the accuracy of the models.

4.4.1.1 Coefficient of determination

R-squared (R^2) is a statistic that explains the amount of variance accounted for in the relationship between two (or more) variables.

The coefficient of determination is given as the ratio of variations explained by the model to the total variations present in Y. Note that the coefficient of determination ranges between 0 and 1. The R^2 value is interpreted as the proportion of variation in Y that is explained by the model. $R^2=1$ indicates that the model exactly explains the variability in Y. On the other hand, $R^2 = 0$ indicates that the model does not explain any variability in Y. R^2 value larger than 0.5 is usually considered a significant relationship.

$$R^2 = 1 - \frac{SSE}{SST}$$

$$SSE = \sum_{i=1}^{n} \left(Y_i - \hat{Y}_i\right)^2 \tag{4.22}$$

$$SST = \sum_{i=1}^{n} \left(Y_i - \overline{Y}_i\right)^2$$

Where SSE and SST represent the error sum of squares and the total sum of squares. These represent the unexplained variance and all variance in the dependent variables. Y_i and $\hat{Y}_i$ are the measured and predicted dependent variables while $\overline{Y}_i$ is the mean of the known dependent variable.

4.4.1.2 Mean absolute error (MAE)

Mean Absolute Error is a model evaluation metric used with regression models. The mean absolute error of a model for a test set is the mean of the absolute values of the individual prediction errors over all instances in the test set. Each prediction error is the difference between the true value and the predicted value for the instance.

The advantages of MAE is that it can deal with data with outliers. The MAE close to zero shows the better model. It is given as:

$$MAE = \frac{1}{n}\sum_{i=1}^{n}|e_i| \quad (4.23)$$

4.5 Anomaly detection

Hawkins (1980), states that an outlier is defined as an observation that deviates so much from other observations as to arouse suspicion that it was generated by a different mechanism. In SHM, detection of outliers or anomalies can be due to damage or a faulty instrument. However, the structural responses are also affected by changing operational and environmental parameters and if not handled carefully, the changes may be assumed to be outliers. Hence the need to follow methodologies that identify outliers, taking the changes in the loads' dams are subject to into consideration.

In dam monitoring, detection of outliers has been determined through comparing current measurements with either the pre-set thresholds obtained from either previous data or finite element models. Other studies have identified outliers by comparing recent observations with those calculated from a reference statistical model. Data collected from the monitoring systems can either be analysed individually i.e., each variable or a combination of variables. Thereafter, univariate, bivariate and multivariate methods are being used in the current practice to detect any anomalies in the data.

Chandola et al., (2009) referred to anomaly detection as the problem of finding patterns in data that do not conform to the expected behaviour. Several methodologies exist in the identification of anomalies in the data. In this study, regression-based techniques will be utilized since Chapter 6 involved the development of statistical models to formulate relationships between structural responses and loads acting on the dam.

Regression-based models involve two steps namely: - (i) fitting a regression model on the data and (ii) use of residuals to determine the anomalies, i.e., the magnitude of the residual is used as an anomaly score (Chandola et.al, 2009). However, the presence of anomalies in the training data can lead to an inaccurate regression model due to anomalies influencing the regression parameters. To eliminate this problem, Rousseuw and Leroy (1987) introduced robust regression which estimates the regression parameters while taking into consideration the anomalies. Also, robust regression can detect anomalies as anomalies have larger residuals from a robust fit. This section concentrates on multivariate methodologies based on robust statistics in the detection of anomalies.

4.5.1 Univariate outlier detection

One of the objectives of this research is to detect any anomalous behaviour in the structural behaviour of Roode Elsberg Dam. In dam engineering, these anomalies used are determined by setting up thresholds for each monitoring instrument installed on the dam. The set thresholds are then used to detect outliers by plotting of univariate control charts (Montgomery, 2001) of residuals obtained from statistical models. Control limits are normally obtained from Equation 4.24 below

$$\begin{aligned} UCL &= \bar{x} + 3\sigma \\ LCL &= \bar{x} - 3\sigma \end{aligned} \tag{4.24}$$

Where UCL= upper control limit, LCL= lower control limit, $\bar{x}$ is the mean and σ is the standard deviation of the residuals.

In this work, thresholds were determined through fitting a distribution to the residuals and choosing values that relate to a percent of the cumulative distribution function.

4.5.2 Bivariate outlier detection

Outliers can also be detected by plotting a scatter plot of two variables. If any point falls outside a predetermined range, then it is said to be an outlier. One way of estimating the outliers is to draw an ellipse on the scatter plot within a specific range (confidence interval). Bagplots proposed by Rousseeuw et al., (1999) are the other bivariate method that can be used to estimate outliers in bivariate data.

4.5.3 Multivariate outlier detection

When dealing with data whose dimension exceeds two, anomaly detection can be done through visual inspection. This makes the process of detecting outliers a difficult process. There exist several methods that have been suggested by different researchers across the globe to detect outliers in multivariate data. Some of these methods to mention but a few include:-

The commonly used method is the so-called Mahalanobis distance method shown in Equation 4.25 (Kannan and Manoj, 2015). This methodology uses the empirical average and covariance matrix as location vector and the shape matrix in Equation 4.25 and tags as an anomaly any observation which has a Mahalanobis squared distance lying above a certain predefined quantile of the distribution with p degrees of freedom (Fauconier and Haesbroeck, 2009).

$$MD_i = \sqrt{((x_i - \mu)C^{-1}(x_i - \mu)^T)} \tag{4.25}$$

Where μ is the sample mean and C is the sample covariance matrix of the data.

One of the shortcomings of Mahalabonis distance method is that it suffers from the so-called masking effect, i.e., one outlier is said to mask another outlier if the second outlier can be considered as an outlier only by itself, but not in the presence of the first outlier (Majewska, 2015).

Multivariate regression analysis techniques are based on least square fitting and this process is affected with the presence of anomalies. If there are anomalies in a data set, the following happens (Hubert et al., 2005):-

- The multivariate estimates differ substantially from the correct answer defined as the estimates obtained in the absence of anomalies
- The resulting fitted model does not allow the detection of the anomalies using their residuals, Mahalanobis distances or the widely used 'leave-one-out' diagnostics.

The above problems can be solved using robust estimators such as the minimum covariate determinant estimator (MCD) suggested by Rousseeuw (1984) to detect anomalies in the data. The MCD estimator is a very robust estimator of multivariate locator and scatters which is computed using the so-called FAST-MCD algorithm developed by Rousseeuw and Van Driessen (1999). Also, for cases of multiple regression, the least trimmed squares (LTS) estimator is used to locate anomalies or outliers.

4.5.3.1 MCD estimator

Rousseeuw (1984) proposed the so-called minimum covariance determinant estimator for determining outliers. This method seeks to find a good fraction h of "good observations" which are not considered to be outliers and compute the sample mean and covariance from this subsample (Leys et al., 2018).

The procedure is repeated for all possible subsamples of size h and at the end the sub-sample which has the minimum determinant is selected. This leads to the deletion of observations that are extreme (outliers) in a robust manner.

The methodology described here is the algorithm described by Gusriani and Firdaniza (2018).

The MCD starts by calculating the subset X as much as h observations given as:

$$h = \frac{(n + k + 1)}{2} \tag{4.26}$$

Then there exists a combination of observations as much as $a = C_h^n$. The subsets of the matrix X are given as:

$$H_b = \begin{bmatrix} x_{11} & x_{12} & \cdots & x_{1k} \\ x_{21} & x_{22} & \cdots & x_{2k} \\ \vdots & \vdots & \ddots & \vdots \\ x_{h1} & x_{h2} & \cdots & x_{hk} \end{bmatrix}, b = 1,2,\ldots, a \tag{4.27}$$

From each H_b, call H_{bm}, the average vector and the covariance matrix are determined by:

$$t_{bm} = \frac{1}{h}(H_{bm})' v \tag{4.28}$$

$$C_{bm} = \frac{1}{h}\left[H_{bm} - vt'_{bm}\right]'\left[H_{bm} - vt'_{bm}\right] \tag{4.29}$$

Where v is the vector column (hx1). Then for each b is calculated the value of $\det(C_{bm})$. For m=1, if $\det(C_{bm}) = 0$, then the distance of the Mahalanobis is calculated in the diagonal of the matrix is determined by:

$$d_{MD}^2 = \left(X^* - v^*\overline{X}^*\right)C^{-1}\left(X^* - v^*\overline{X}^*\right)' \tag{4.30}$$

4.6 Chapter summary

Dam health monitoring deals with the collection of data from continuous monitoring systems installed on dams. This data can only be used when correctly interpreted such that correct decisions can be made concerning the structural health of dams. Gaussian process regression is a machine learning approach that can be used to build statistical models from monitoring data. To make proper decisions concerning the status of a dam, it is important to detect any anomalies in the data collected from the monitoring schemes. In this chapter, the fundamentals of two regression methodologies namely: Gaussian process regression and least trimmed squares were presented. The important steps in GPR, i.e. choice of kernel functions and hyperparameters were described herein. The concept behind the least trimmed squares in the regression was also presented in this chapter.

Also, performance indicators, namely RMSE, MAE and R2 used in the comparison of model performance were described in detail. Outlier detection is an important step in dam monitoring, therefore methods such as control charts and least trimmed square-based methods were described in this chapter. In the next chapter, continuous monitoring schemes at Roode Elsberg Dam as a case study are described.

5 Continuous monitoring of Roode Elsberg Dam

5.1 Introduction

The majority of dams around the world are more than 50 years old. Therefore, the dam engineers need to ensure that the safety of these ageing dams is handled properly as the risk associated with them is of low probability but of high consequence. To ensure the safety of dams, there is a need to collect all data on these ageing dams through field measurements such that if there are any deviations from previous data, immediate action can be taken to avoid failure.

In addition to field measurements, dam safety can be also be achieved through continuous dam monitoring to identify as soon as possible any abnormalities in the structural behaviour of dams such that forewarning can be allowed for and remedial works can be done before a disaster occurs (Pang et al., 2006). Dam monitoring comprises monitoring of loads (hydrostatic pressure, temperature etc.) and corresponding structural responses that include static responses (stresses, strains, deformations, etc.) and dynamic properties (natural frequencies, damping ratios and mode shapes). In dam engineering, the common practice is deformation monitoring which comprises of instruments that measure a few displacements at chosen points on the dam structures and foundations (Frodl and Naterop, 2007). The commonly used type of monitoring is deformation monitoring of dams (crest, foundation movements) carried out either by TRIVEC instruments or now geodetic surveys or global positioning systems described in Chapter Two. In recent years, dam engineers and researchers have embarked on incorporating dynamic behaviour of dams in continuous monitoring through ambient vibration monitoring (Bukenya et al., 2014, Oliviera et al., 2007). The basis for ambient vibration monitoring is that dynamic properties of a dam are related to its stiffness and water level. The dynamic properties of a dam are expected to remain the same at a given water level; any deviations would indicate anomalous behaviour.

In South Africa, continuous dam monitoring has been going on for the last decade including deformation monitoring and recently, dynamic monitoring has commenced. Therefore, this chapter aims at describing the deformation and dynamic monitoring systems installed on a South African dam (Roode Elsberg Dam). Observations from these monitoring schemes are also presented to understand the variation of the structural behaviour of the dam over time and the interdependencies between the loads and dam responses. In this chapter, four data sets are described namely: - (i) reservoir water level, (ii) air temperature, (iii) water temperatures (iv) deformations and (v) modes of vibration. The monitoring data was provided by both the Department of Water and Sanitation and the University of Cape Town.

5.2 Description of Roode Elsberg Dam

Roode Elsberg Dam (Figures 5.1 and 5.2) is a double curvature concrete arch dam with a centrally situated overspill section. The height above the foundation is 72m, the length of the crest is 274m and the gross capacity of the reservoir is 8.210 million m^3. Table 5.1 shows the details of Roode Elsberg Dam. Roode Elsberg Dam was constructed in 1968 situated on the Sand Drift River 30km from Worcester, Western Cape Province, South Africa. The dam was constructed to provide supplementary water to establish a more reliable supply of irrigation water to the farms in the Hex River Valley.

Figure 5.1. Roode Elsberg Dam

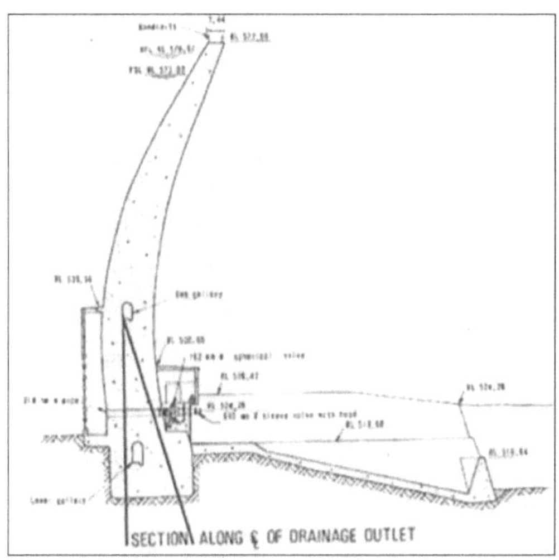

Figure 5.2. Cross-section of Roode Elsberg Dam

Table 5.1. Main characteristics of Roode Elsberg Dam

Foundation maximum height	72 m
Crest width	2.6 m
Crest length	274 m
Spillway length	76 m
Reservoir normal capacity	8.21 x10^6 m^3
Reservoir minimum capacity	2.05 x10^6 m^3

5.3 Continuous dam monitoring systems

The use of monitoring systems on a dam site is to measure loads acting on the dam and the dam structural responses resulting from the loads. There exist two continuous monitoring systems installed on Roode Elsberg Dam namely: - Deformation and ambient vibration monitoring systems. Table 5.2 shows the suite of instruments installed at Roode Elsberg Dam for continuous monitoring of the dam.

Table 5.2. Instruments used in Roode Elsberg Dam

Instrumentation	**Parameter measured**
Weather station	Air temperature
GNNS	Deformations
Pressure transmitter	Water level
Accelerometer	Natural frequencies
Thermal couples	Water temperature

5.3.1 Deformation monitoring system

Deformation monitoring system uses Global Navigation Satellite Systems (GNSS) such as Global Position System (GPS). The GNNS based technique is used to measure geodetic networks in obtaining precise point displacements in key locations on and around dams (Van Cranenbroeck, 2011). At Roode Elsberg Dam, GNNS was installed on the dam in 2010 to measure dam deformations continuously. The system consists of four Trimble GNSS sensors with two primary sensors (P01 and P03) at key locations near the dam and two secondary sensors (P203 and P206) located on the dam crest as shown in Figure 5.3. Figure 5.4 shows the GPS receiver installed on the dam crest. Data collected from these sensors are transferred to the server located in the control room. The server runs Trimble 4D software which helps in the computation of deformation of the dam at an hourly basis.

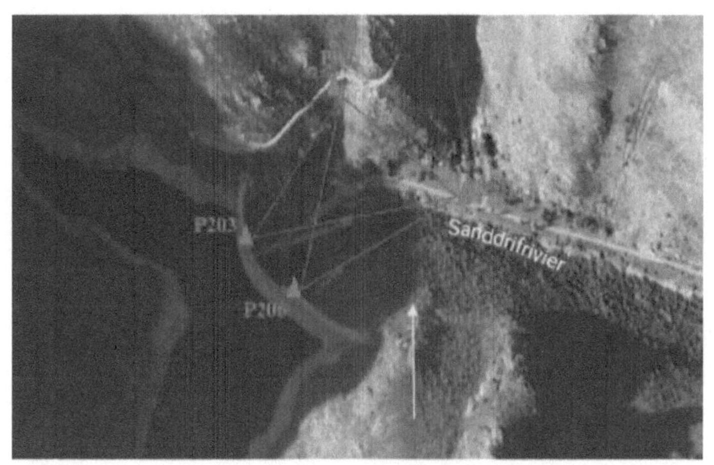

Figure 5.3. GPS network at Roode Elsberg Dam

Figure 5.4. GPS receiver on the dam crest

5.4 Ambient vibration monitoring

Ambient vibration monitoring (AVM) of dams facilitates the determination of dynamic properties (natural frequencies, mode shapes and damping ratios). Changes in the dynamic properties of dams indicate changes in the physical properties such as stiffness and boundary conditions of the structure. AVM is not a common practice in dam engineering as most reported studies (Kemp, 1996, Gaftoi et al., 2016) focused on once-off ambient vibration tests.

In South Africa, ambient vibration monitoring of dams started with the carrying out of biannual (summer and winter) ambient vibration tests on Roode Elsberg Dam since 2008. Section 5.4.1 below describes the ambient vibration tests that have been performed on Roode Elsberg Dam before the installation of the continuous dynamic monitoring system.

5.4.1 Ambient vibration tests

Several ambient vibration surveys were conducted on Roode Elsberg Dam between 2008 and 2013. The main objectives of these tests were to (i) update the finite element model and (ii) investigate the effect of changes in the natural frequencies of Roode Elsberg Dam caused by changes in reservoir water level.

5.4.1.1 Experimental setup

During the ambient vibration tests performed on Roode Elsberg Dam, eight 10V/g tri-axial forced balanced accelerometers (Figure 5.5) were placed on eight blocks of the dam crest equalling to a total of 24 accelerometers. Figure 5.6 shows the layout of the field tests performed on the dam. The twenty-four accelerometers were then connected to specialised data acquisition hardware manufactured by National Instruments with channels that can simultaneously acquire each channel with 24-bit high-resolution delta-sigma analogue-to-digital converters (ADCs). The system has anti-aliasing filters to prevent aliasing and noise from affecting the measurement quality. The accelerometers were connected to NI PXI-4472B device cards mounted into a NI PXI 1045 chassis which was connected via fibre cable to a PCI card of an ordinary PC workstation.

Two of the eight units were fixed at reference points (7 and 8) whereas the other six were placed at the remaining points on the dam crest measuring the accelerations in three (radial, tangential and vertical) orthogonal directions. For each test, the time series of 30 minutes were collected at a sampling frequency of 1000 Hz.

Figure 5.5. Tri-axial forced balanced accelerometers placed on the dam crest

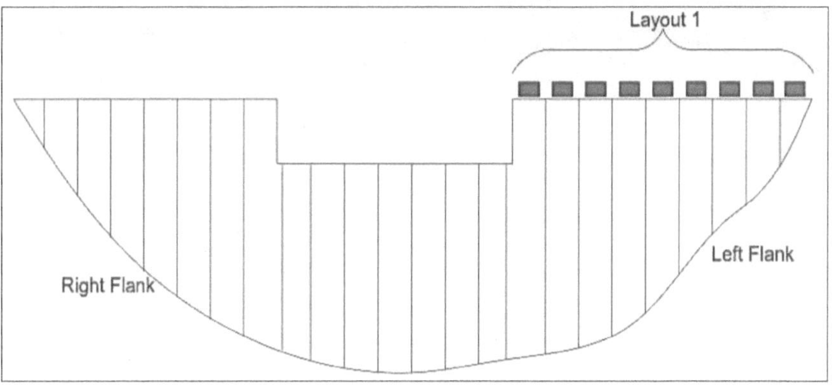

Figure 5.6. Test set up

5.4.1.2 Estimation of natural frequencies

To estimate natural frequencies of Roode Elsberg Dam, modal parameter estimation techniques had to be applied to the acceleration time series (Figure 5.7) measured during the field tests. The process involves: -

(i) data preprocessing, i.e., detrending of the data to remove any offsets, low pass filtering and decimation to reduce the sampling frequency from 1000Hz to 20 Hz.

(ii) data post-processing, i.e. application of OMA techniques to the pre-processed data to extract natural frequencies. Two OMA techniques were applied namely: -the frequency domain decomposition (FDD) technique in the frequency domain and stochastic subspace identification (SSI) technique in the time domain. These techniques are available in commercial software called ARTeMIS (SVS,2010).

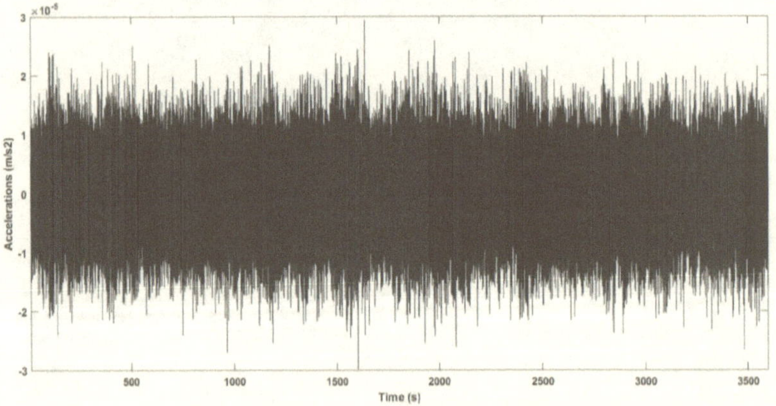

Figure 5.7. Sample of acceleration time series

The application of FDD and SSI techniques on the ambient vibration data allowed the identification of 3 stable modes in the frequency range of 3- 6.5Hz. Singular values obtained from using the FDD technique are shown in Figure 5.8 whereas the stabilization diagram resulting from using the SSI-method is shown in Figure 5.9.

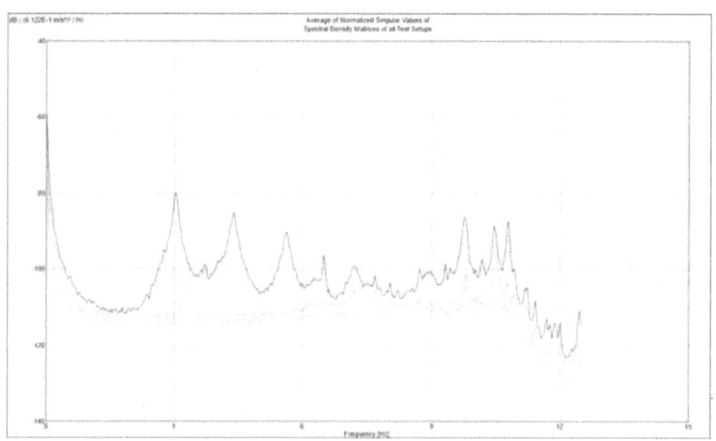

Figure 5.8. Sample of Spectral density

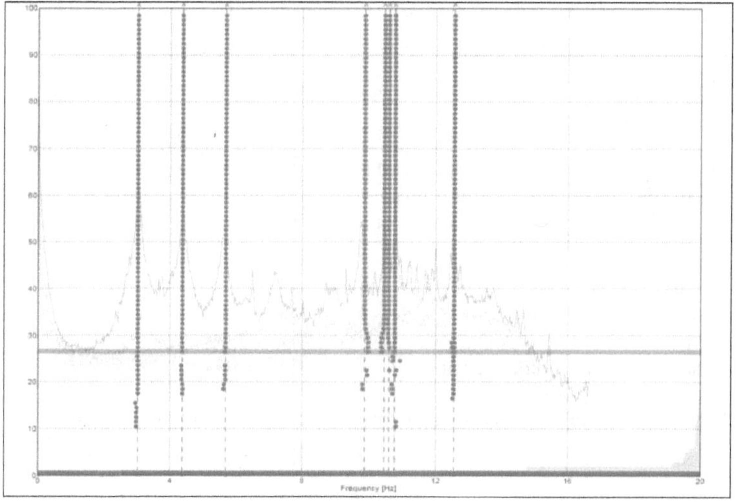

Figure 5.9. Stabilization diagram from the SSI method

Table 5.3 shows the modes that were estimated from the six ambient vibration measurements on Roode Elsberg Dam.

Table 5.3. Natural frequencies from different tests

Dec 2008	April 2009	Sept 2009	April 2010	Dec 2010	Sept 2013
3.09	3.46	3.00	3.50	3.05	3.02
4.39	4.84	4.43	4.85	4.44	4.42
5.64	6.02	5.64	6.03	5.64	5.59

The following observations can be made from the ambient vibration tests:

(i) Three modes were determined in the frequency range of 3-6.5Hz.

(ii) Natural frequencies of dams increased from September corresponding to the period when the dam is full to April when the dam levels are low every year.

(iii) Results from the ambient vibration measurements provided evidence that AVT is an effective tool for assessing the structural condition of dams and, at the same time, suggested the continuous dynamic monitoring strategy as a suitable strategy for SHM.

Section 5.4.2 below describes the ambient vibration monitoring system installed on Roode Elsberg Dam.

5.4.2 Description of the ambient vibration monitoring system

A simple permanent ambient vibration monitoring system was installed on the dam in December 2013 to monitor the evolution of the dynamic properties of the dam over time under different environmental conditions.

The ambient vibration monitoring system installed on Roode Elsberg Dam consists of three tri-axial accelerometers installed on the dam crest (Figures 5.10 and 5.11). The accelerometers were placed on the dam blocks that had been most active during the previous periodic ambient vibration tests. The accelerometers have a dynamic range of more than 125dB at ± 2g full scale within the 0.1 to 50Hz range. The accelerometers are connected to the digitizer which communicates with the industrial computer running the GeoDAS software using an Ethernet cable. The computer and the digitizer are placed on the dam wall inside the upper gallery of the dam to avoid damage due to any adverse weather conditions on the dam site. The system collects and stores acceleration time series at a sampling rate of 50Hz after every hour.

Figure 5.10. The ambient vibration monitoring system

● Accelerometers

Figure 5.11. Position of accelerometers on the dam crest

To estimate the dynamic properties of the dam, the collected acceleration time series are processed by:-

- Pre-processing of data, i.e., elimination of the trend, filtering using Hanning window
- Estimation of natural frequencies using SSI-data method using ARTeMIS Extractor Pro 4.5.

5.5 Measurement of operational and environmental factors

Dams are subjected mainly to two loads namely, hydrostatic and thermal loads. The hydrostatic load is normally measured through water level and thermal load through temperature (air, water and dam body). This section presents how these loads are measured at Roode Elsberg Dam.

5.5.1 Reservoir water level

For dams, it is important to monitor the water level in the reservoir regularly to determine the quantity of water in the reservoir and its level relative to the regular outlet works and the emergency spillway. Reservoir water level helps in the computation of hydrostatic pressure and pore pressure; the volume of seepage is usually directly related to the reservoir water level. It is also important to establish the normal or typical flow through the outlet works for legal purposes. The water level at Roode Elsberg Dam is measured by a submersible pressure transmitter daily.

5.5.2 Air and water temperature

5.5.2.1 Air temperature

Temperature measuring devices are very important in arch dams since volume changes caused by temperature fluctuations have a significant contribution to the loading on an arch dam (USACE, 1994). Thermometers are used to determine the temperature gradients for use in evaluating thermal stresses, which contribute to thermal cracking. Temperature measurements are important both to determine causes of movement due to expansion or contraction and to compute actual movement. To ensure that there are temperature measurements taken at Roode Elsberg Dam, a weather station was installed on the crest of the dam (Figure 5.12).

Figure 5.12. Weather station situated on the dam crest

5.5.2.2 Reservoir water temperature

Some thermometers are embedded in the dam wall to measure reservoir water temperatures at different levels. Figure 5.13 shows a schematic view of the position of thermometers that are installed on the dam wall at different levels to measure water temperatures.

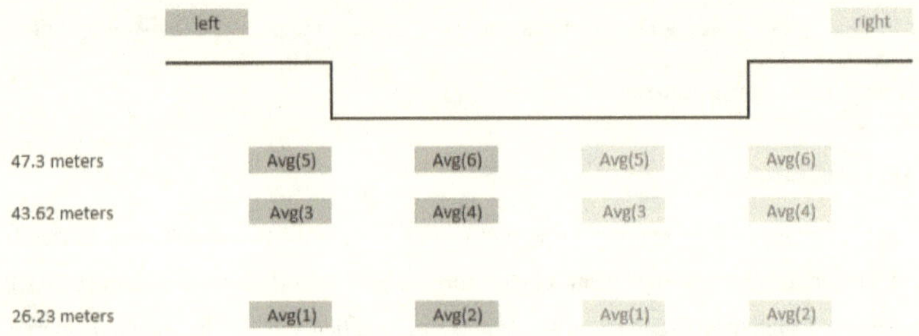

Figure 5.13. Position of thermometers (Avg) measuring water temperatures

5.6 Chapter summary

Continuous monitoring of dams is very important in ensuring dam safety since any deviations from normal dam behaviour can be detected on time. Two monitoring systems (deformation and ambient vibration) installed at Roode Elsberg Dam have been described in this chapter. Furthermore, before the installation of the ambient vibration monitoring system at the dam, several ambient vibration tests were conducted out on the dam. Results showed that changes in natural frequencies were due mainly to change in water level. Changes in dam natural frequencies due to temperature could not be ascertained as the field tests were only performed in winter and summer. Chapter Six discusses observations made from the monitoring data collected from the two continuous monitoring systems. Results obtained from the formulation of relationships between loads (temperature and water level) and dam responses (deformations and natural frequencies) through statistical modelling are discussed herein.

6 Results and discussions

6.1 Introduction

Continuous monitoring of Roode Elsberg Dam has been in operation since 2001 for deformation monitoring and 25th November 2013 for dynamic monitoring. This implies that there is a lot of data available to understand the structural behaviour of the dam. In this chapter, results obtained from the first 2 years of operation are presented. The purposes of this chapter were first to present observations from the collected monitoring data, presented such that (i) the evolution of both loads (water level and temperature) on the dam structure and the structural responses (deformations and natural frequencies) are obtained (ii) the effect of environmental and operational factors on both deformation and natural frequencies can be understood. Secondly, to present results obtained from the prediction of natural frequencies and deformations using GPR models and performance comparison with other models. Lastly, results from anomaly detection methodologies are explored.

6.2 Observations from monitoring data

The amount of monitoring data collected from civil engineering structures such as dams cannot be useful unless this data is analysed carefully. The information collected from the two continuous monitoring systems described in Chapter 5 allows us to develop structural health monitoring strategies in dam safety. This is done through understanding the effect of changing operational and environmental conditions on the structural response (deformations and natural frequencies) of the dam. This section discusses observations made from the results of the two monitoring schemes installed at Roode Elsberg Dam.

6.2.1 Reservoir water level

The reservoir water levels of Roode Elsberg Dam are being monitored daily since the construction of the dam. Figure 6.1 shows the variation of water level from February 2012 to June 2017. Observation of the water level indicates that there is a regular annual cyclic pattern with maximum and minimum water levels occurring in June to August and Jan to May respectively. The maximum and minimum water levels measured during this period were 48.996m in August 2013 and 18.319m in May 2016 respectively.

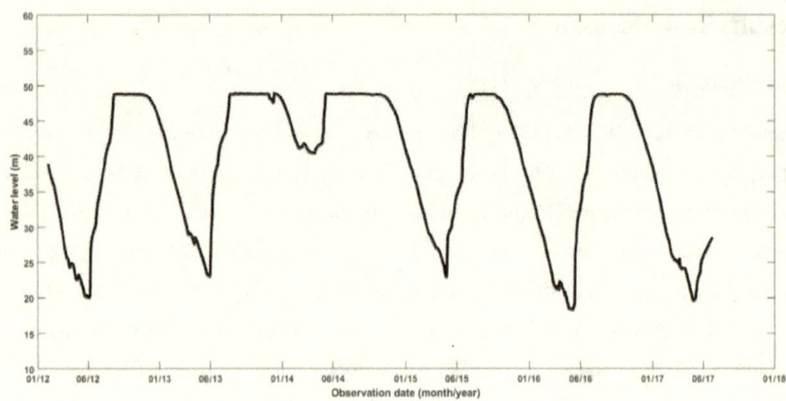

Figure 6.1. Time series of water level at Roode Elsberg Dam

6.2.2 Air and water temperatures

Figure 6.2 shows the variation of daily maximum, daily average and daily minimum temperatures recorded at Roode Elsberg Dam site for a period between 01/01/2012 to 30/06/2017. Figure 6.2 shows that maximum and minimum temperatures occur in February and July respectively. The maximum and minimum daily recorded temperatures at the dam site was 42.2 ° C on 16th January 2012 and 9th July 2014.

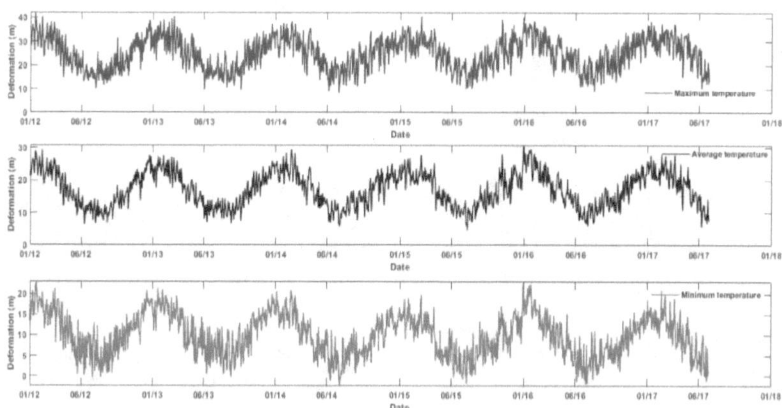

Figure 6.2. Time series of daily maximum, average and minimum air temperature at Roode Elsberg Dam

Figure 6.3 shows the variation of water temperatures measured at different levels (Figure 5.13) at Roode Elsberg Dam during the period of 01/01/2012 to 30/06/2017. WT1 and WT2 are temperatures corresponding to the thermocouples at 47.3m, and WT3 and WT4 correspond to thermocouples at 43.62m. WT5 and WT6 represent temperatures recorded by thermocouples at 26.23m. Figure 6.3 shows that water temperatures WT3 to WT6 are more stable because most of the time these were not exposed to direct sunshine during the measuring period.

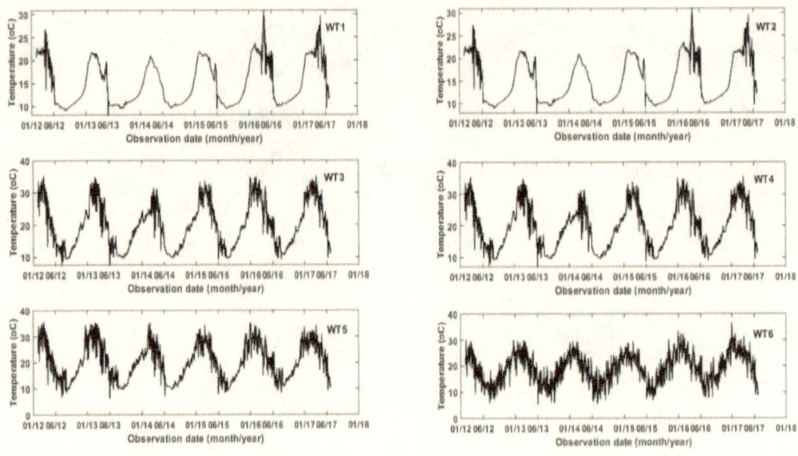

Figure 6.3. Time series of water temperatures at Roode Elsberg Dam

6.2.3 Natural frequencies of Roode Elsberg Dam

6.2.3.1 Observations

Natural frequencies of Roode Elsberg Dam were extracted from acceleration time-series recorded by the ambient vibration monitoring system. Data-driven stochastic subspace identification (SSI-data) method is used to estimate the natural frequencies of the dam. SSI-data method is one of the time-based output-only modal analysis methods that utilize stabilization diagrams to estimate the modal parameters (Zhang et al., 2005). This technique has been shown to perform better than other out-put only techniques such as frequency domain techniques (Bukenya, 2011). Figure 6.4 shows a typical stabilization diagram used in the estimation of the natural frequencies using the SSI-data method with model orders between 5 and 100. The first three natural frequencies were identified and their modal properties are shown in Table 6.1. The results presented in Table 6.1 correspond to the average of each of the values identified from the three accelerometers.

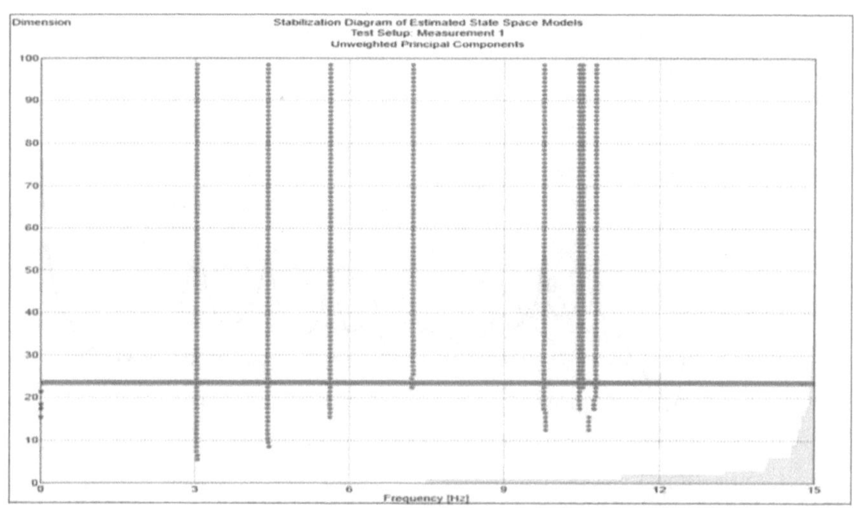

Figure 6.4. Stabilization diagram from SSI-data method

Table 6.1. First three natural frequencies of Roode Elsberg Dam

Mode	f_{mean} **[Hz]**	***Description***
1	3.261	Asymmetric
2	4.664	symmetric
3	5.896	symmetric

Figure 6.5 shows that the time series of three natural frequencies extracted from acceleration data from the dam represents a 24-hour average. The blank spaces represent periods when there was a system failure and during this period there is no data available. The identification of natural frequencies from data sets collected during the period from December 2013 to February 2017

provided the frequency tracking of the first three modes as shown in Figure 6.5 below. Figure 6.5 reveals the following observations: -

(i) There is a seasonal variation of natural frequencies caused by the environmental loads namely water level and temperature. The three modes in the range 3Hz-6Hz being monitored increases from January to around May; this is attributed to the reduction in water level and an increase in the temperature. At the beginning of June, there is a big drop in the natural frequencies because the water level of the dam increased owing to winter rains, then natural frequencies remain constant until November. The modes then start to increase in December and the cycle continues. It should be noted that there is a large scatter in the modes in February and May as this was the period of low water level. This may be attributed to the independent movements of the blocks where the sensors were installed.

(ii) The sensitivity of natural frequencies to low reservoir water levels is seen from February to May during the period of analysis.

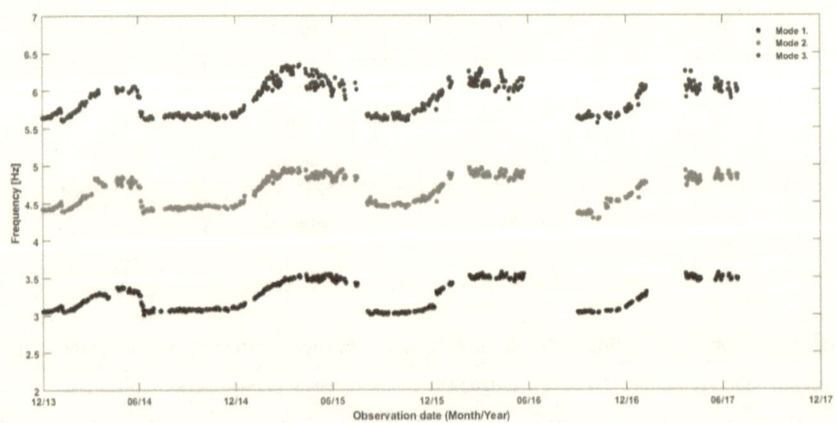

Figure 6.5.Time evolution of the first three natural frequencies of Roode Elsberg Dam

6.2.3.2 Interactions between natural frequencies and environmental variables

To understand the relationship between natural frequencies and the environmental variables, plots of the two variables are plotted against each other. Figure 6.5 shows a plot of water level versus the first natural frequency. Figure 6.5 shows that the rise in water level leads to a decrease in natural frequency. This can be explained using Equation 6.1 below.

$$\omega = \sqrt{\frac{K}{M + m}} \tag{6.1}$$

where ω represents the natural frequency, K represents the stiffness of the dam, M is the mass of the dam wall and m is the mass of water behind the dam wall. From Equation 6.1 K and M are constant and the only variable is the mass of the water behind the dam wall. The mass of the water changes with time and when it increases, natural frequencies decrease since the denominator of Equation 6.1 increases.

To further understand the interaction, time series of water level and mode 1 are plotted on the same graph as shown in Figure 6.6 below. Figure 6.6 reveals that between A and B, the water level remains constant but there is an increase in natural frequency (line EF) at a rate of 0.0002; the same observation is observed in region DE. Region BC shows a decrease in water level causing an increase in the natural frequency.

To investigate the effect of air temperature on natural frequencies, a plot of air temperature versus the first natural frequency is shown in Figure 6.8. The influence of temperature variation on natural frequencies of the dam is not obvious from a comparison of temperature data to natural frequency data. Also, regions AB and CD on Figure 6.8 represent periods of high and low water levels respectively. However, a closer observation of the variation of modes with time in the period, June 2014 to November 2014 shows a slight increase in the natural frequency with the water level almost constant. The average monthly temperature increased during this period (see Figure 6.9) and thus, it can be concluded that temperature change leads to a change in natural frequency.

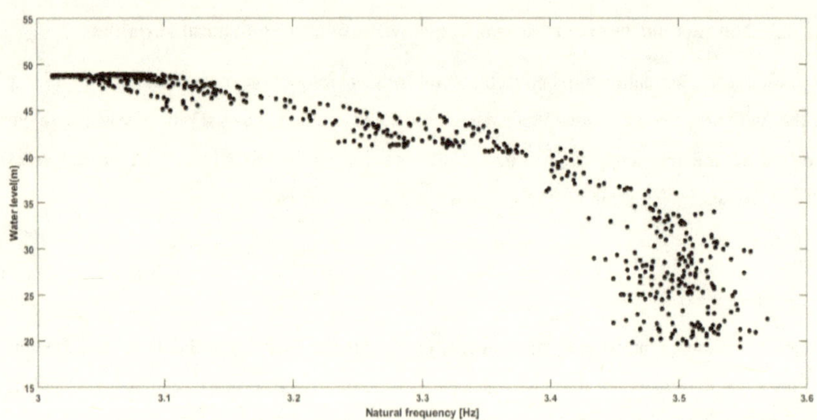

Figure 6.6. Relationship between water level and the first natural frequency

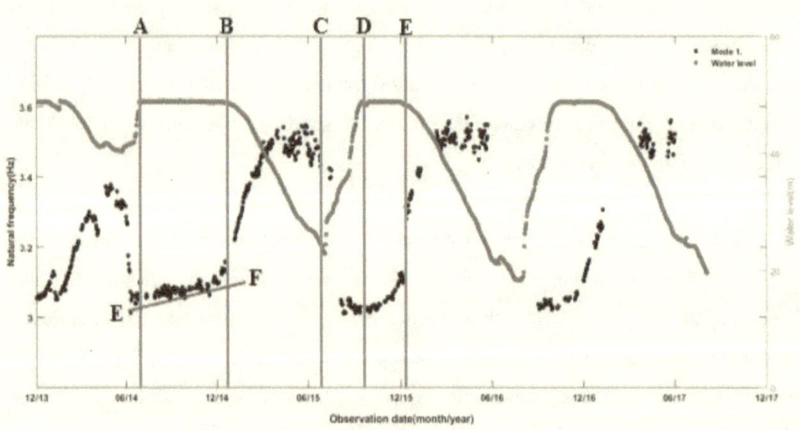

Figure 6.7. Evolution of water level and first natural frequency

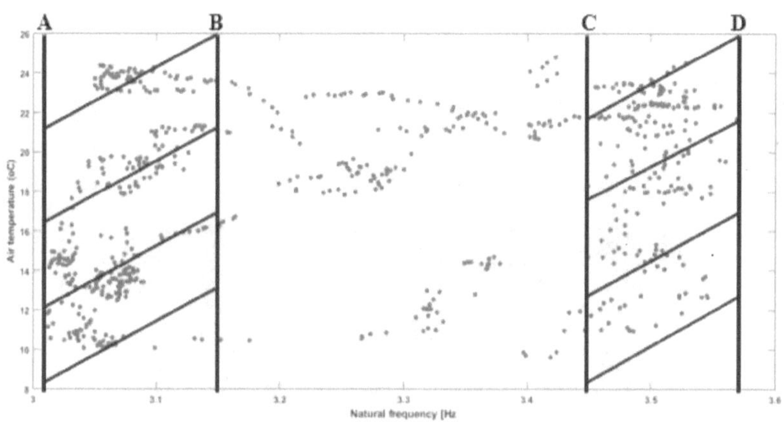

Figure 6.8. Relationship between air temperature and first natural frequency

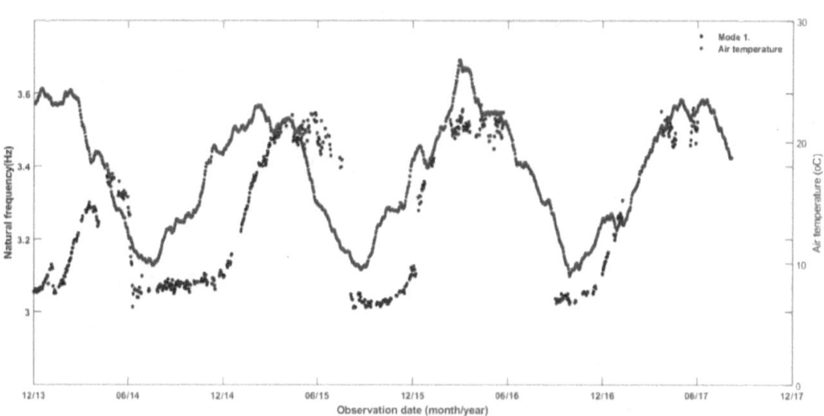

Figure 6.9. Evolution of air temperature and natural frequency

6.2.4 Deformations of Roode Elsberg Dam

6.2.4.1 Observations

Deformations were measured by the GNNS system at Roode Elsberg Dam. In this study, radial deformations, tangential deformations and vertical deformations were measured in the directions Y, X and Z respectively as shown in Figure 6.10 below.

Figure 6.10. Directions of measured dam deformations

The data available for the present study is from February 2012 to February 2016. Table 6.2 below shows statistics of the deformation measured from Roode Elsberg Dam indicating the minimum mean, and maximum values. The absolute maximum radial deformation was about 40mm radially which occurred in the winter season in July. This is expected as it is a period when the dam was full, i.e., 48.787 m which is equivalent to 7.748mil m^3 of water and behind the dam wall which caused the whole body to move downstream. Table 6.2 further shows that the dam deforms more on the right-hand side than on the left due to the different geological conditions, i.e., the right-hand side is softer than the left-hand side concerning rock conditions.

Table 6.2. Descriptive statistics of Roode Elsberg Dam deformations

Variable	Minimum (mm)	Mean (mm)	Maximum (mm)
dy1	0	15.1	39.9
dx1	0	3.5	10.1
dz1	0	8.7	22
dy2	0	16.6	39.9
dx2	0	3.9	10.9
dz2	0	8.2	19

Figures 6.11-6.13 show plots of radial, tangential and vertical deformations against time recorded on Roode Elsberg Dam respectively. Positive deformations in radial direction indicate downstream movement and the negative displacements represent upstream movements. In the tangential direction, positive and negative deformation represent east and west movements respectively while in a vertical direction, they represent up and down movements respectively. Gaps in the graph represent periods when the system was faulty.

Figure 6.11 shows the radial deformations measured on the dam from February 2012 to February 2016. The positive and negative values of the radial deformations represent upstream deformations and downstream deformations respectively.

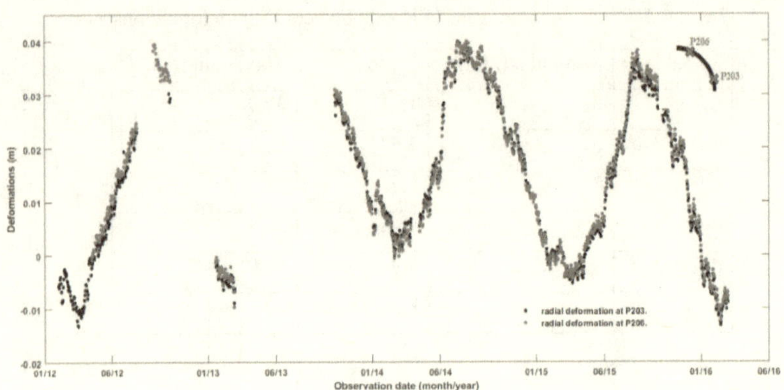

Figure 6.11. Radial deformations at Roode Elsberg Dam

Figure 6.12 shows tangential deformations measured on Roode Elsberg Dam. The positive and negative values of the tangential deformations represent outward and inward movements of the dam in the x-axis respectively.

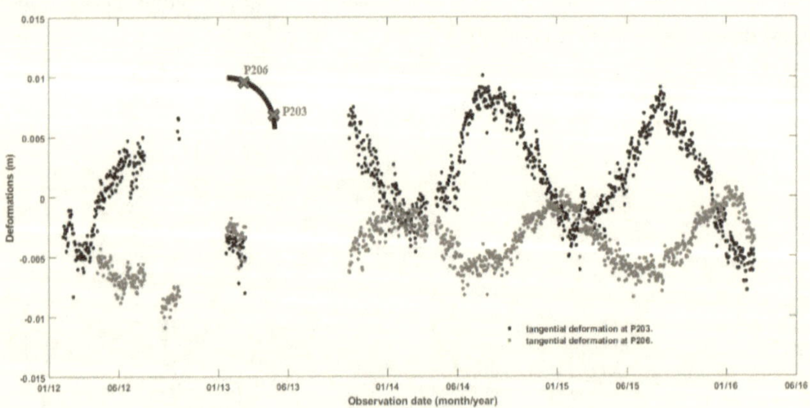

Figure 6.12. Tangential deformations at Roode Elsberg Dam

Figure 6.13 shows vertical deformations measured on Roode Elsberg Dam. The positive and negative values of the vertical deformations represent an increase and decrease in height respectively.

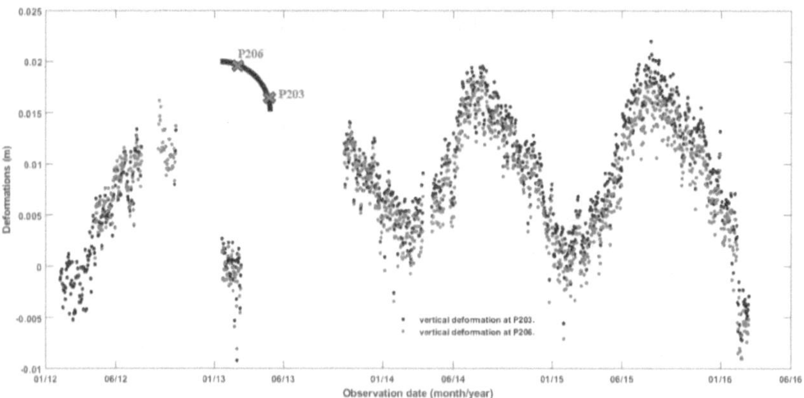

Figure 6.13. Vertical deformations at Roode Elsberg Dam

6.2.4.2 Interactions between deformations and environmental variables

To understand the relationship between deformations and the environmental variables, plots of the variables were made. i.e water level versus deformations and temperature versus deformations.

Radial deformations

Figure 6.14 shows a plot of reservoir water level versus radial deformation between 2014 to 2015. A similar pattern seen in Figure 6.14 was observed for all the years. It is interesting to note that in 2012 when the dam level was around 40m, the dam deformed 10mm downstream and it was the same deformation in 2012 and 2015 at lower water levels of 20m. To understand further the interaction between water level and radial deformations, both factors were plotted on the same plot. Figure 6.15 shows that radial deformations of Roode Elsberg Dam are influenced by changes in reservoir water level. It can be observed that when the water level rises, the dam wall is displaced downstream due to the added mass by the rising water behind the dam wall. This continues until the dam reaches a deformation of 40mm when the dam is spilling. At full capacity, the dam wall starts to move upstream and this period corresponds to the temperature rising due to the beginning of the summer period in the area (region 1 in Figure 6.14). When the water level starts to go down, the dam keeps on moving back to its un-deformed position due to the removal of the added water mass (region 2 in Figure 6.14). Then as the water level rises, the dam wall moves downstream to maximum deformation of around 40mm (region 3 in Figure 6.14).

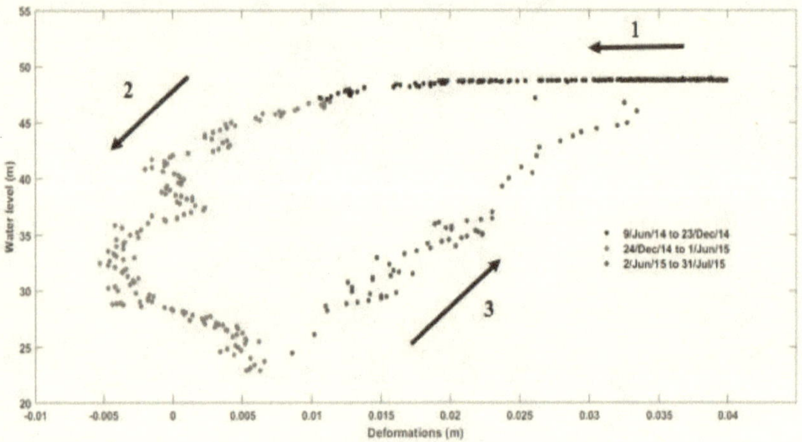

Figure 6.14. Water level vs radial deformations

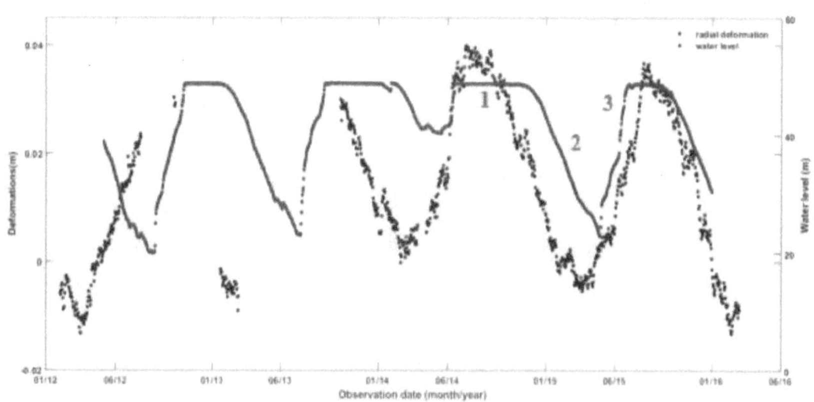

Figure 6.15. Evolution of water level and radial deformations

Figure 6.16 shows a plot of air temperature versus radial deformations. It can be observed that there is a negative correlation between air temperature and radial deformations, i.e., when temperature increases, the dam moves towards downstream and when temperature decreases, the dam moves towards upstream (regions 1 to 3 in Figure 6.16). This is also explained in Figure 6.17 which shows a time series of air temperature and radial deformation.

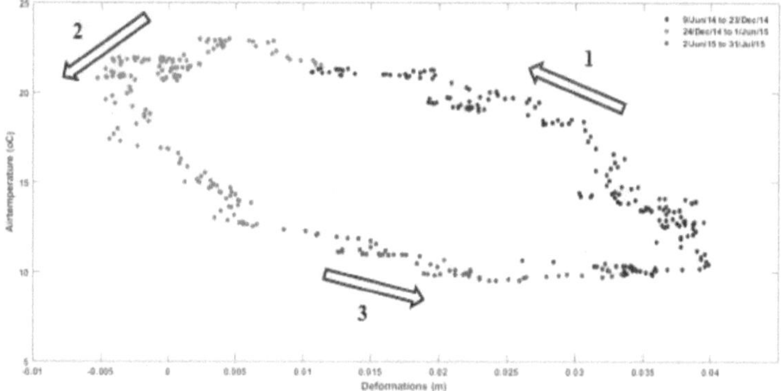

Figure 6.16. Air temperature vs radial deformations

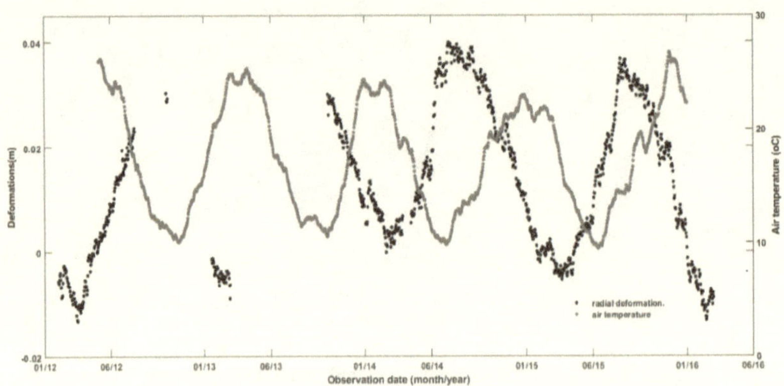

Figure 6.17. Evolution of air temperature and radial deformations

To understand the effect of water temperatures on radial deformations, time series of both radial deformations water temperatures at 26.53m (see Figure 5.13) were plotted as shown in Figure 6.18. Figure 6.18 shows that when water temperatures are lowest the dam is moving downstream and as the temperature increases, the dam moves downstream. This trend is similar to that exhibited by temperature versus radial deformation.

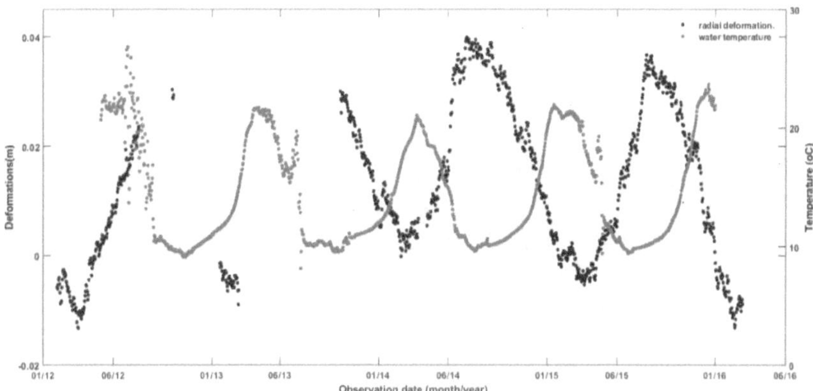

Figure 6.18. Time series of water temperature and radial deformations

Tangential deformations

Figure 6.19 shows a plot of reservoir water level versus tangential deformation. Observation of this plot shows that when the dam levels are high the dam is pushed to opposite directions for both the right flank (P206) and left flank (P203). This is because the arch is expanding and it moves further on the right flank of the arch than on the left flank.

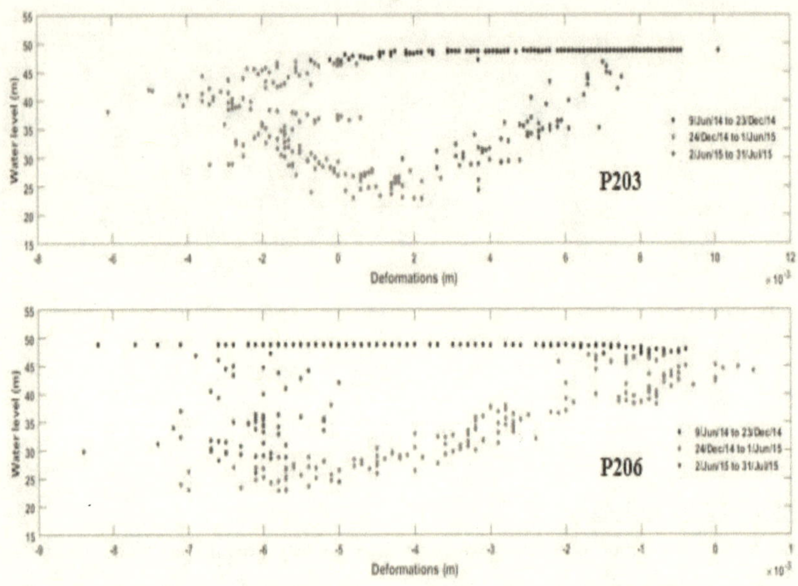

Figure 6.19. Water level vs tangential deformations

Figure 6.20 shows a plot of air temperature and tangential deformations. It can be observed that temperature significantly affects tangential deformations. The tangential deformations on the left flank of the dam have a negative correlation with temperature as when temperature goes up, the deformation goes down and vice versa. On the other hand, the tangential deformation has a positive correlation with temperature, i.e., as temperature goes up, the deformation goes up.

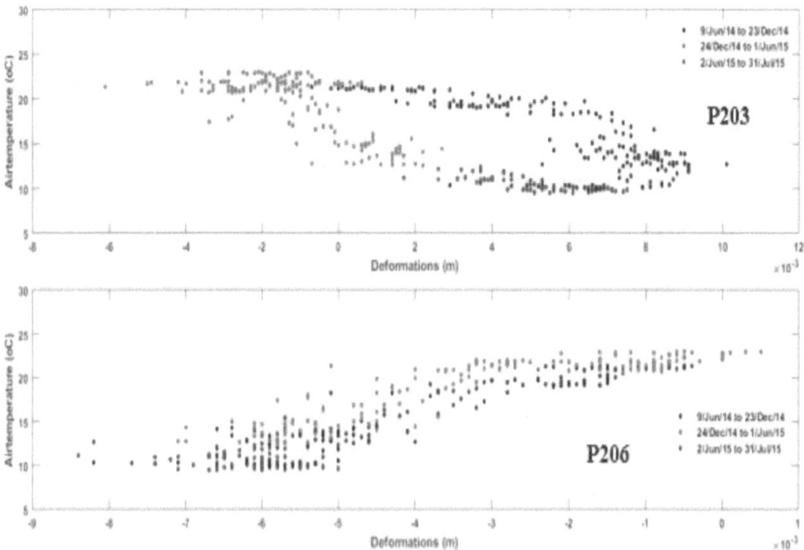

Figure 6.20. Air temperature vs tangential deformations

Vertical deformations

Figure 6.21 shows plots of reservoir water level versus vertical deformations. Although there is no clear pattern as the scatter is big when the water level is high the dam settles down and when the water level is low, there is an uplift of the dam wall.

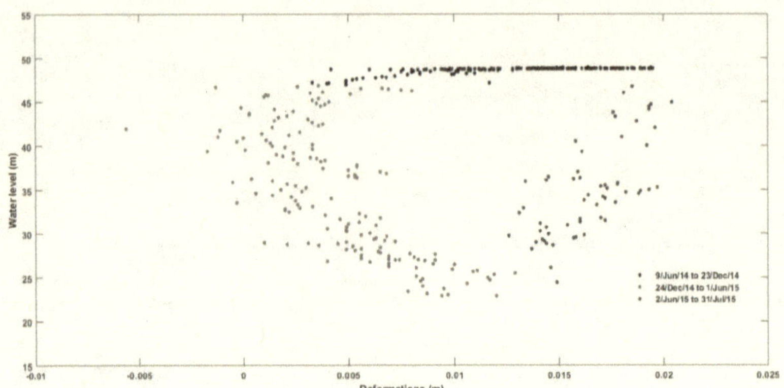

Figure 6.21. Water level vs vertical deformations

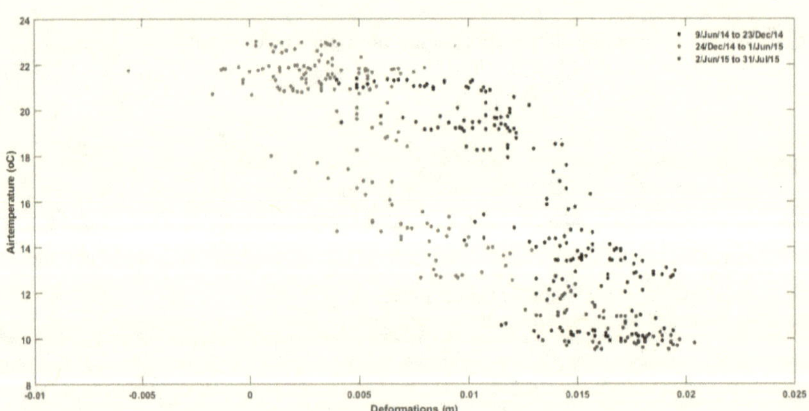

Figure 6.22. Air temperature vs vertical deformations

6.2.5 Relationship between dam deformations and dynamic monitoring

The relationship between deformations and natural frequencies can be realised using the equation of motion:

$$(M + m)\ddot{x} + C\dot{x} + K(x + \Delta x) = F \tag{6.2}$$

Where M is the mass of the wall,

m is the added mass due to water behind the dam wall

x is the deformation

$\ddot{x}$ and $\dot{x}$ is the acceleration and velocity respectively

K, F is the stiffness and force respectively

$\ddot{x}$ is related to the natural frequency of a structure,

Δx is directly proportional to change in temperature of the structure, i.e., $\Delta x \alpha \Delta T$.

Static and dynamic monitoring systems are installed on Roode Elsberg Dam. Before the installation of the continuous dynamic monitoring system on the dam, several full-scale ambient vibration tests were conducted on the dam. These ambient vibration tests made it possible to define mode shapes of the dam. The following observations were made relating dam deformations and mode shapes.

- It was also observed that left flank and right flank deformed in opposite directions in the tangential direction. This movement in the tangential direction corresponds to the first mode shape of the dam (Figure 6.23).

- In the radial direction, the wall deforms towards downstream and then as the water level decreases, the wall moves upstream due to the changes in water level and temperature. This movement in the radial direction corresponds to the second mode shape of the dam (Figure 6.24).

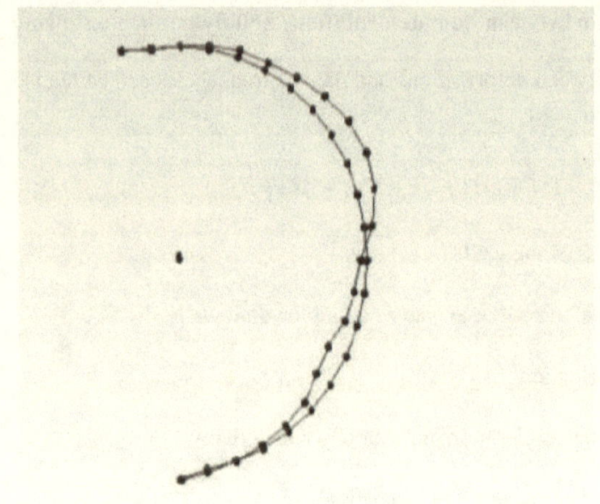

Figure 6.23. First mode shape

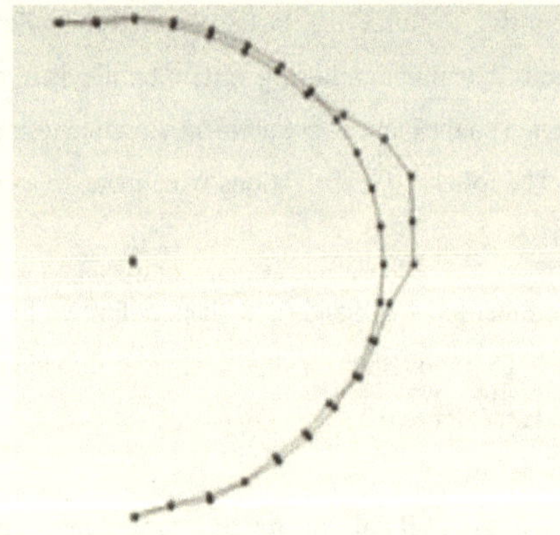

Figure 6.24. Second mode shape

6.2.6 Summary of observations

- The natural frequencies of the dam are strongly dependent on the reservoir level, i.e., the natural frequencies of the dam decrease with increase in water level.
- The influence of temperature on natural frequency is clear during the periods when the water level was constant. In these periods, the increase in temperature leads to an increase in the natural frequency.
- The maximum radial deformations of Roode Elsberg Dam is 40 mm.
- The maximum vertical deformations of Roode Elsberg Dam are about 2 times smaller than the maximum radial deformations.
- The maximum tangential and vertical deformations of Roode Elsberg Dam are 10mm and 22mm respectively.
- Tangential deformations of Roode Elsberg Dam move in opposite directions due to different ground conditions on either side of the spillway.
- The seasonal deformations of the dam are strongly dependent on temperature variation. While the seasonal variation of deformations is largely driven by temperature variation, the reservoir level does influence the deformations of the dam especially as the dam fills rapidly.

In conclusion, observations from the monitoring data collected from the monitoring schemes installed on Roode Elsberg Dam have shown that changes in both deformations and natural frequencies are due to changes in water level and temperature (air and water). To understand these relationships, there is a need to develop statistical models that will enable us to predict the dam responses and also detect any abnormal behaviour. Section 6.3 discusses results obtained from statistical modelling of data collected from the monitoring systems installed on Roode Elsberg Dam.

6.3 Statistical modelling of dam natural frequencies and deformations

6.3.1 Introduction

Changes in dam responses (natural frequencies and deformations) may indicate damage or deterioration of the structure. However, dams are also subjected to operational and environmental parameters (water level, temperature, wind, ground motion). These parameters can also cause a change in the dam responses. Therefore, it is important to understand the changes in dam responses due to environmental and operational parameters through statistical modelling between dam responses and environmental and operational measured data. Figure 6.25 shows methodologies that were used in the prediction of dam responses using temperature and water level as environmental and operational factors.

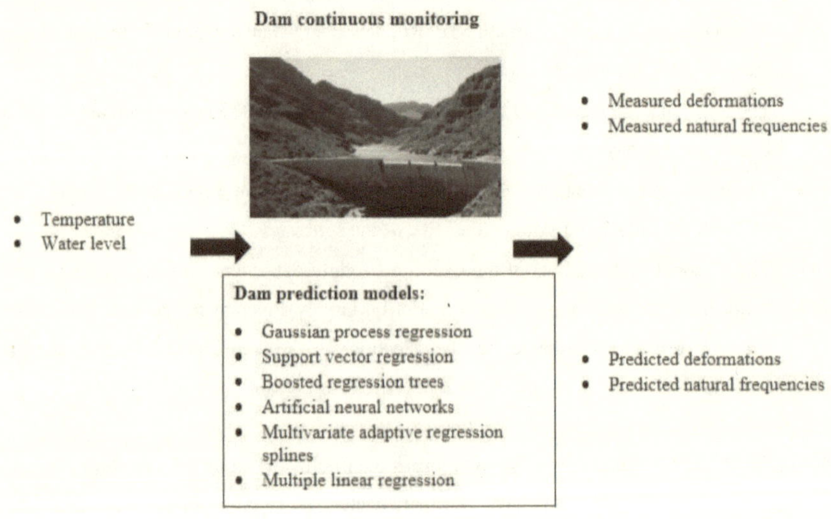

Figure 6.25. Prediction methodologies used in statistical modelling of dam monitoring data

The following steps were followed in the analysis of data collected from the monitoring systems at Roode Elsberg Dam. The analysis of data was implemented in Matlab.

Step 1: Correlation analysis to determine how variables are correlated. The variables that are highly correlated were used in the regression analysis.

Step 2: Normalise the selected data set from the monitoring data between 0 and 1.

Step 3: Divide the data into training and test data sets.

Step 4: Construct the selected statistical models shown in Figure 6.23 using the training data set.in particular, for GPR modelling, where different covariance functions are tested such that the best covariance function for dam monitoring data is chosen.

Step 5: Test the models constructed in step 4 with the test data set such that the general performance of the models can be obtained. Using performance indicators, the performance of the different models is compared.

Step 6. Outlier detection methodology is carried out to ascertain if the dam was behaving normally during the period of analysis.

6.3.2 Canonical correlation analysis of monitoring data

To study the relationship between the structural responses (deformation and natural frequencies) and environmental conditions (air temperature and water level), canonical correlation analysis was carried out. Table 6.3 summarises the canonical analysis results. The largest canonical correlation value was 0.9837 which indicated that there is a strong influence that the environmental and operational parameters (V1) have on the structural responses of the dam (U1).

Table 6.3. Canonical correlation of structural responses versus environmental parameters

Canonical Correlation	Wilks' lambda	χ^2	Degree of Freedom	p-value
0.9837	0.0017	4631.5	18	0
0.9741	0.0512	2148.7	8	0

Furthermore, a correlation study was carried out between the structural responses and environmental parameters. Table 6.4 shows the variables with the highest correlation coefficient. Observations show that deformations have a strong correlation with both water level and temperature while natural frequencies are strongly correlated to water level.

Table 6.4: Strongly correlated parameters

Variable	Correlation
dy1-airtemp	0.877
dx1-airtemp	0.923

dz1-air temp	0.904
dy2-airtemp	0.885
dz2-air temp	0.901
Freq1-wl	-0.939
Freq2-wl	-0.892
Freq3-wl	-0.868

Figure 6.26 shows a scatter plot of U1 and V1 canonical variables which shows that there is an existing linear relationship between the variables. This plot shows that environmental and operational parameters can be used to predict the structural response of dams.

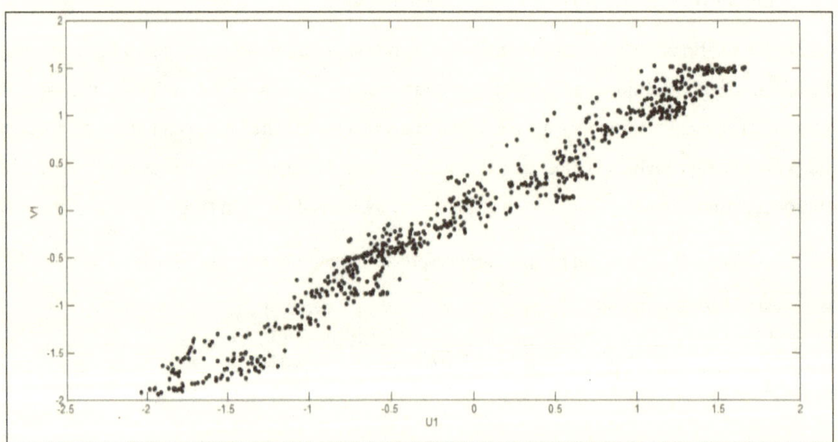

Figure 6.26. Plot of canonical variables U1 and V1

In summary, the canonical correlation analysis of data collected from Roode Elsberg Dam has exhibited that loads and dam responses are related. Section 6.3.3 presents results obtained from the statistical modelling of natural frequencies and deformations collected from Roode Elsberg Dam.

6.3.3 Natural frequency-based prediction models

In this section, the effectiveness of GPR-based models using different kernel functions in the prediction of natural frequencies using temperature and water level collected from Roode Elsberg Dam is discussed. Dam monitoring data collected from December 2013 to June 2017 on Roode Elsberg Dam were divided into two sets namely: - training and testing set in a ratio of 70:30. Results obtained from GPR-based models are compared with other data-driven models such as SVR, ANN, MARS, BRT and MLR.

6.3.3.1 Gaussian Process Regression

In this study, various GPR models based on the following kernel functions, that is, Exponential kernel, Squared Exponential kernel, Rational Quadratic kernel and Matern 5/2 kernel were used in the prediction of dam natural frequencies using temperature and water level. To assess the quality of the predictions using R^2 value, the root mean squared error (RMSE), and the mean absolute error (MAE) were used. The values of the error indicators are shown in Table 6.5. Table 6.5 presents the prediction performance of GPR models with different kernel functions for the first natural frequency. As seen in Table 6.5, the GPR model based on Exp kernel has the best performance in the training stage having the lowest RMSE value of 0.077. In the testing stage, the R.Quad-based GPR model had the best performance with the lowest RMSE value of 0.278. Overall, the R.Quad-based GPR model had the best performance among the kernels. Figure 6.27 shows the scatter plots of observed values and predicted values of the first natural frequency using GPR models with different kernel functions. Figure 6.27 indicates that values are distributed around the centre line and are in agreement with the measured data.

Table 6.5. Prediction accuracy of 1st natural frequency based on GPR models using different kernels

Kernel	Training			Testing		
	R^2	MAE	RMSE	R^2	MAE	RMSE
SqE	0.99	0.059	0.083	0.86	0.256	0.372
Matern5/2	0.99	0.055	0.078	0.86	0.241	0.374
Exp	0.99	0.052	0.077	0.77	0.204	0.471
R.Quad	0.99	0.055	0.078	0.92	0.223	0.278

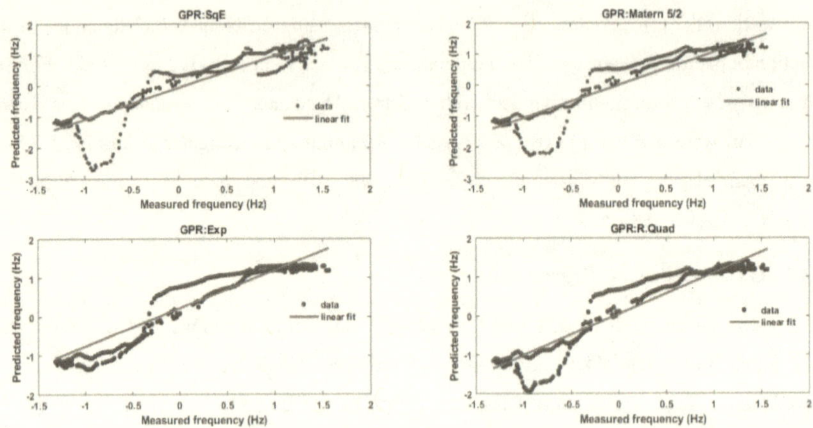

Figure 6.27. Observed frequency vs predicted 1st natural frequency using GPR models

Table 6.6 presents the prediction performance of GPR models with different kernel functions for the second natural frequency. It can be observed that the R.Quad-based GPR model has the best performance in the training stage having the lowest RMSE value of 0.107. In the testing stage, R-Quad based GPR model had the best performance with the lowest RMSE value of 0.336. Overall, the R-Quad-based GPR model had the best performance among the kernels.

Table 6.6. Prediction accuracy of 2nd natural frequency based on GPR models using different kernels

Kernel	Training			Testing		
	R^2	MAE	RMSE	R^2	MAE	RMSE
SqE	0.99	0.074	0.111	0.86	0.317	0.366
Matern5/2	0.99	0.070	0.108	0.88	0.294	0.339
Exp	0.99	0.066	0.109	0.85	0.333	0.378
R.Quad	0.99	0.068	0.107	0.86	0.323	0.366

Table 6.7 presents the prediction performance of GPR models with different kernel functions for the third natural frequency. It can be observed that the GPR model based on Exp kernel has the

best performance in the training stage having the lowest RMSE value of 0.147. In the testing stage, R Quad-kernel had the best performance with the lowest RMSE values of 0.362 respectively. Overall, the R Quad-based GPR model had the best performance among the kernels.

Table 6.7. Prediction accuracy of 3rd natural frequency based on GPR models using different kernels

Kernel	Training			Testing		
	R^2	MAE	RMSE	R^2	MAE	RMSE
SqE	0.98	0.102	0.156	0.81	0.299	0.391
Matern5/2	0.98	0.095	0.152	0.82	0.300	0.381
Exp	0.98	0.089	0.147	0.84	0.297	0.366
R.Quad	0.98	0.091	0.148	0.84	0.364	0.362

One of the objectives of this research study was to understand the effect of water temperature on natural frequencies. Similar procedures were followed in the prediction of natural frequencies using GPR models with water level and water temperatures as inputs in the regression analysis. Results obtained in the analysis are shown in Appendices A1 and A2. Results show that the prediction accuracy reduced by 30% for the natural frequencies respectively. This showed that water temperatures do not have a significant effect on natural frequencies.

To investigate the overall performance of the different kernel functions, GPR models were developed for the natural frequencies using the entire data set. Table 6.8 presents the prediction performance of GPR models with different kernel functions for the first natural frequency using the entire data set. Results showed that the GPR model based on Exp kernel has the best performance with the lowest MAE and RMSE values of 0.054 and 0.081 respectively. Tables 6.8 and 6.9 show that Exp-based GPR models had the lowest MAE and RMSE values in the prediction of second and third natural frequencies respectively.

Figure 6.28 shows the measured and predicted first natural frequency using the GPR model with Exponential kernel function with the corresponding residuals.

Table 6.8. Prediction accuracy of GPR models using the whole data set for the first natural frequency

Kernel	Criteria		
	R^2	MAE	RMSE
SqE	0.99	0.063	0.089
Matern5/2	0.99	0.059	0.085
Exp	0.99	0.054	0.081
R.Quad	0.99	0.056	0.082

Table 6.9. Prediction accuracy of GPR models using the whole data set for second natural frequency

Kernel	Criteria		
	R^2	MAE	RMSE
SqE	0.97	0.11335	0.17622
Matern5/2	0.97	0.10749	0.17169
Exp	0.97	0.10052	0.16616
R.Quad	0.97	0.10341	0.16912

Table 6.10. Prediction accuracy of GPR models using the whole data set for third natural frequency

Kernel	Criteria		
	R^2	MAE	RMSE
SqE	0.97	0.10824	0.16795
Matern5/2	0.97	0.0987	0.15938
Exp	0.97	0.09586	0.15817
R.Quad	0.98	0.95234	0.15679

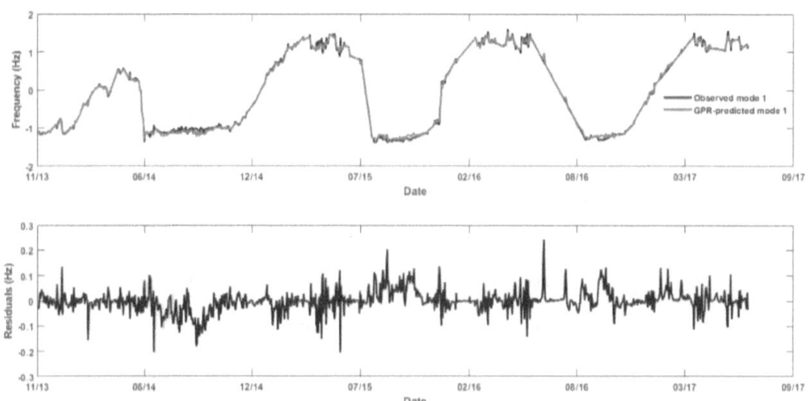

Figure 6.28. Observed and predicted first natural frequency using GPR-Exp model

6.3.3.2 Performance of other machine learning natural frequency models

Tables 6.11, 6.12 and 6.13 show the performance parameters namely R-squared, mean absolute error and root mean square error for selected machine learning models (SVR, BRT and ANN) for training and test data set for the first, second and third natural frequencies of Roode Elsberg Dam respectively. Results show that SVR performed better than ANN and BRT with the lowest RMSE value in the prediction of natural frequencies.

Table 6.11. Performance of ML models on 1st natural frequency

Model	Training			Testing		
	R^2	MAE	RMSE	R^2	MAE	RMSE
SVR	0.99	0.08	0.106	0.93	0.193	0.243
BRT	0.99	0.079	0.100	0.93	0.206	0.278
ANN	0.99	0.081	0.102	0.88	0.272	0.386

Table 6.12. Performance of ML models on 2nd natural frequency

Model	Training			Testing		
	R^2	MAE	RMSE	R^2	MAE	RMSE
SVR	0.98	0.102	0.133	0.87	0.309	0.315
BRT	0.98	0.112	0.148	0.82	0.353	0.383
ANN	0.98	0.101	**0.138**	0.84	0.340	0.385

Table 6.13. Performance of ML models on 3rd natural frequency

Model	Training			Testing		
	R^2	MAE	RMSE	R^2	MAE	RMSE
SVR	0.97	0.118	0.166	0.86	0.253	0.338
BRT	0.96	0.148	0.211	0.83	0.289	0.395
ANN	0.97	0.137	0.191	0.86	0.342	0.396

The overall prediction performance of machine learning models was tested on the entire set of natural frequencies. Tables 6.13, 6.14 and 6.15 show the performance of these models in the prediction of first, second and third natural frequencies respectively. Results indicate that SVR outperforms BRT and ANN because it has the lowest RMSE value. Figures 6.29, 6.30 and 6.31 show plots of observed and predicted values of the first natural frequency using BRT, SVR and ANN models respectively. These figures show that SVR predictions are close to the observed values.

Table 6.14. Prediction accuracy of ML models using the whole data set for the first natural frequency

Kernel	Criteria		
	R^2	MAE	RMSE
BRT	0.98	0.1113	0.1382
SVR	0.99	0.1117	0.0931
ANN	0.99	0.0864	0.1098

Table 6.15. Prediction accuracy of ML models using the whole data set for second natural frequency

Kernel	Criteria		
	R^2	MAE	RMSE
BRT	0.95	0.1649	0.2267

SVR	0.97	0.1264	0.1844
ANN	0.96	0.1303	0.1845

Table 6.16. Prediction accuracy of ML models using the entire data set for third natural frequency

Kernel	Criteria		
	R^2	MAE	RMSE
BRT	0.95	0.166	0.232
SVR	0.97	0.128	0.183
ANN	0.96	0.139	0.191

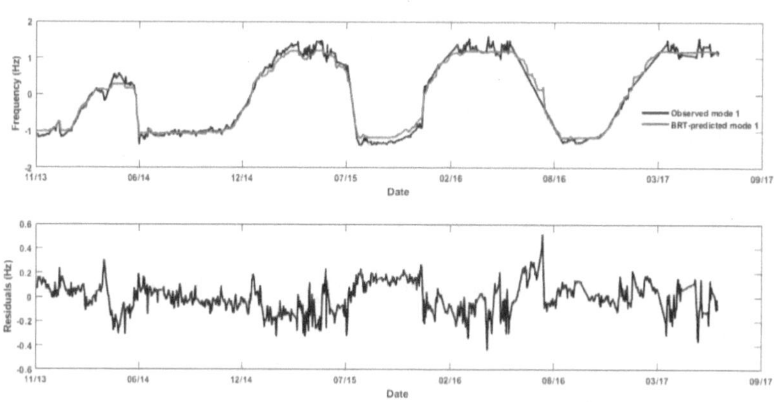

Figure 6.29. Observed and predicted first natural frequency using the BRT model

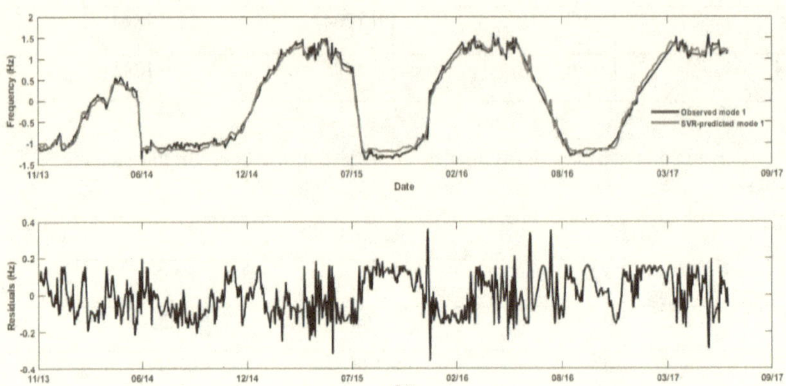

Figure 6.30. Observed and predicted first natural frequency using the SVR model

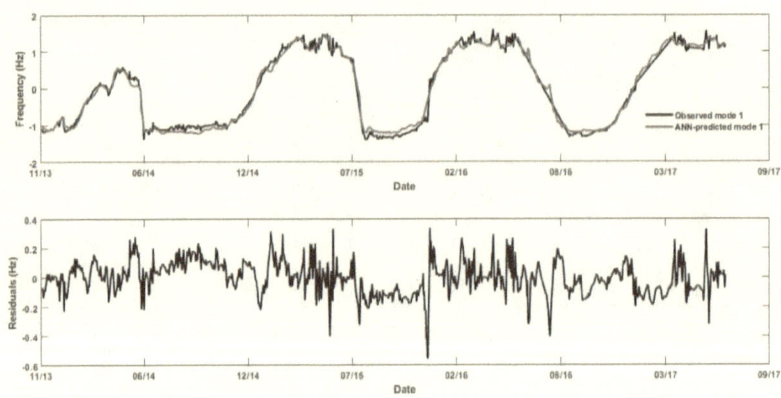

Figure 6.31. Observed and predicted first natural frequency using the ANN model

6.3.3.3 Multivariate adaptive regression splines

Data was split into training and testing tests in the ratio of 70% and 30 % respectively. In this analysis, the piecewise cubic model was used in the modelling of natural frequencies as they provided a better approximation than the linear models. During the process of modelling using MARS, a set of BFs are selected by forward and backward iterations. Then the partitioned points are compared using the RSME in the iterations. Finally, the so-called modified generalised cross-validation (GCV) is used as a criterion to choose a final model. Also, MARS uses ANOVA to generate and compare models in terms of the importance of the inputs. Table 6.17 below shows the ANOVA decomposition selected using the GCV as a criterion for radial functions. The GCV column represents the significance of the corresponding ANOVA function by giving the GCV value for a model with all BFs corresponding to that particular ANOVA function to be calculated. The performance of MARS models is presented in Table 6.18.

Table 6.17. ANOVA decomposition of the MARS model selected with GCV criterion

Function	Std	GCV	#Basis	#Params	Variable
1	0.839	0.0476	1	2.5	1
2	29.109	0.0145	3	7.5	2

Figure 6.25 shows a plot of how GCVs and MSE values change during the iterations. This figure shows that the lowest value of GCV occurs at 20 BFs. The same criterion was used for tangential and vertical deformations.

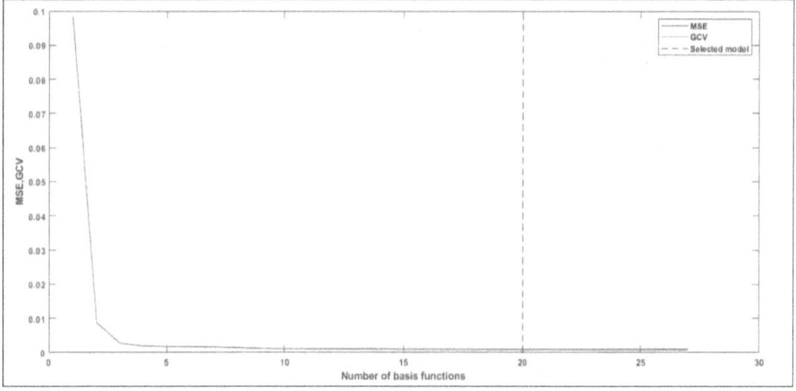

Figure 6.32. MARS model parameters for natural frequency

Table 6.18. Performance of MARS developed models for natural frequency

	Training			Test		
Target/Criteria	R^2	MAE	RMSE	R^2	MAE	RMSE
Mode 1	0.99	0.077	0.101	0.94	0.217	0.264
Mode 2	0.98	0.100	0.137	0.83	0.334	0.391
Mode 3	0.97	0.134	0.189	0..82	0.335	0.431

To get the overall performance of MARS models, the entire data set of natural frequencies were subjected to MARS models. Table 6.19 presents the performance of the models on the first, second and third natural frequencies. Figure 6.33 shows the time series of observed and predicted first natural frequency with corresponding residues due to MARS models.

Table 6.19. Prediction accuracy of MARS models using the entire data set for all natural frequencies

Modes	Criteria		
	R^2	MAE	RMSE
Mode 1	0.98	0.111	0.142
Mode 2	0.95	0..152	0.211
Mode 3	0.94	0.172	0.238

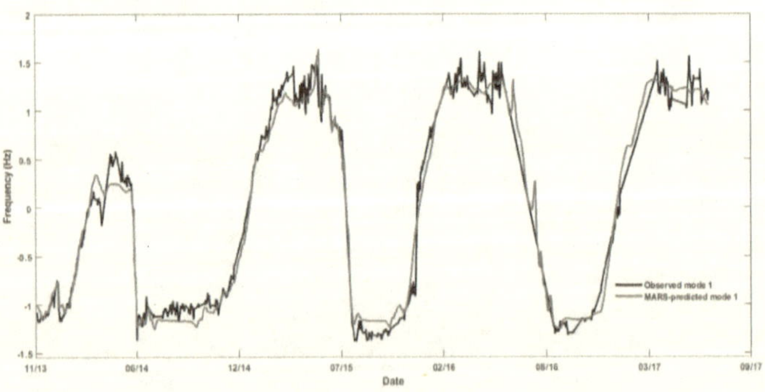

Figure 6.33. Measured and predicted mode 1 using the MARS model

6.3.3.4 Multiple Linear regression

In modelling of natural frequencies from Roode Elsberg Dam, the known HST model described in Chapter 3 was used. In this study, only the effects due to hydrostatic and thermal loads were used in the model due to the length of the data being monitored not being long enough. Due to the lack of temperatures of the concrete dam wall, the thermal load was estimated using sinusoidal functions as seen in Equation 3.1 The time effect was not considered due to the shorter time used time series of the measured deformations, December 2013 to June 2017. Using the least-squares method, the best MLR model was obtained using Equation 3.1. Table 6.20 shows the obtained regression coefficients for the first three natural frequencies. Table 6.21 shows the performance of MLR models on the natural frequencies for both training and testing data set.

Table 6.20. Performance of MLR developed models for natural frequencies

Parameter		Mode 1	Mode 2	Mode 3
Regression coefficients	β_0	3.161	4.506	5.737
	β_1			
	β_2	1.8884	4.7569	4.1135
	β_3	-1.6667	-8.5862	-6.6341
	β_4		4.1102	2.6752
	β_5	0.0350		0.0492
	β_6	-0.08235	-0.0822	-0.0867
	β_7	-0.0561		
	β_8	0.0386		-0.0251

Table 6.21. Performance of MLR models on natural frequencies

Target/Criteria	Training			Testing		
	R^2	MAE	RMSE	R^2	MAE	RMSE
Mode 1	0.87	0.0475	0.0671	0.85	0.0401	0.080
Mode 2	0.81	0.0590	0.0843	0.78	0.0656	0.087

Mode 3	0.85	0.060	0.0833	0.83	0.060	0.086

Table 6.22 shows the performance of MLR models using the entire data set for all natural frequencies. Results show that the MLR model can be used in modelling of natural frequencies. Figure 6.34 shows the measured and predicted values of the first natural frequency using the MLR model.

Table 6.22. Prediction accuracy of MLR models using the entire data set for all natural frequencies

Modes	Criteria		
	R^2	MAE	RMSE
Mode 1	0.90	0.042	0.059
Mode 2	0.84	0.056	0.078
Mode 3	0.85	0.058	0.079

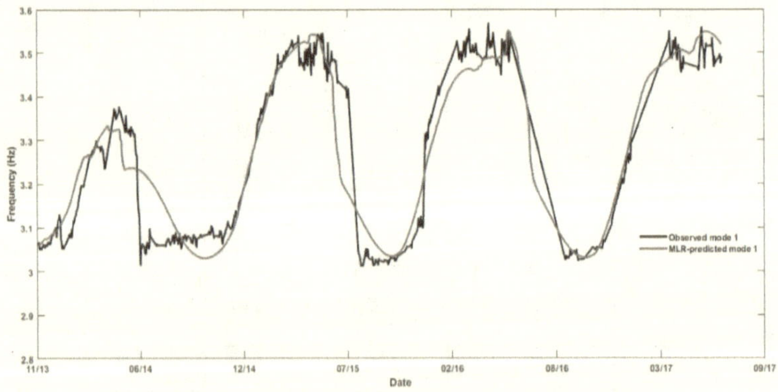

Figure 6.34.Measured and predicted mode 1 using MLR model

To differentiate the effect of water level and temperature on natural frequencies, MLR models were applied to the entire data set. Table 6.22 shows the performance of MLR models in the prediction of the first, second and third natural frequencies. Figure 6.35 shows the effect of water level and temperature as predicted by the MLR model on the first natural frequency. Figure 6.35 shows that water level is a dominant factor in the variation of natural frequencies.

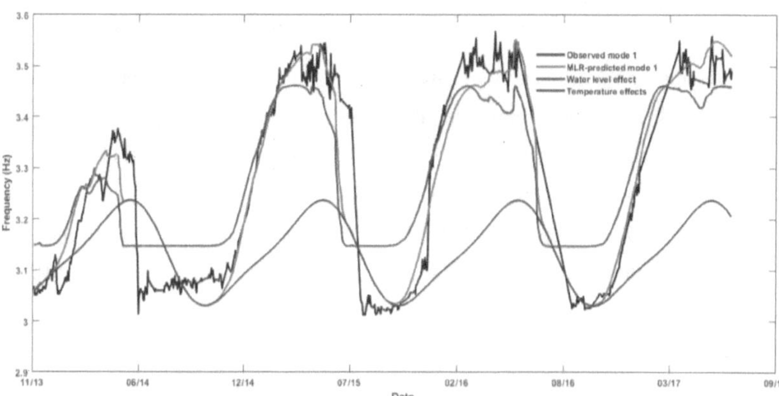

Figure 6.35. First natural frequency time series of measured and fitted values and its components by hydraulic and temperature model

In summary, statistical modelling of natural frequencies is an important step in understanding the dynamic behaviour of dams. Results have shown that GPR models performed better than other models in the modelling of natural frequencies. Generally, machine learning models performed better than the MARS and MLR models.

6.3.4 Deformation based prediction models

In this section, the effectiveness of GPR-based models using different kernel functions in the prediction of deformations using temperature and water level collected from Roode Elsberg Dam is discussed. Dam monitoring data collected from February 2012 to February 2016 (1491 observations) on Roode Elsberg Dam was divided into two sets namely: - training and testing set in a ratio of 70:30. Results obtained from GPR-based models are compared with other data-driven models such as SVR, ANN, MARS, BRT and MLR.

6.3.4.1 Gaussian Process Regression

In this study, various GPR models based on the following kernel functions, that is, Exponential kernel, Squared Exponential kernel, Rational Quadratic kernel and Matern 5/2 kernel were used in the prediction of dam deformations using temperature and water level.

Table 6.23 presents the prediction performance of GPR models with different kernel functions for radial deformations. It can be observed that the GPR model based on Exp kernel has the best performance in the training stage having the lowest MAE and RMSE values of 0.1225 and 0.1862 respectively. In the testing stage, R.Quad-based GPR model had the best performance with the lowest MAE and RMSE values of 0.1978 and 0.2533 respectively. Overall, R.Quad-based GPR model had the best performance among the kernels. Figure 6.34 shows the scatter plots of observed values and predicted values of the radial deformations using GPR models. Figure 6.34 indicates that values are distributed around the centre line.

Table 6.23. Prediction accuracy of radial deformation based on GPR models using different kernels

Kernel	Training			Testing		
	R^2	MAE	RMSE	R^2	MAE	RMSE
SqE	0.96	0.1347	0.1968	0.87	0.2625	0.3475
Matern5/2	0.96	0.1321	0.1949	0.90	0.2325	0.3098
Exp	0.96	0.1225	0.1862	0.93	0.2273	0.2677
R.Quad	0.96	0.1262	0.1919	0.93	0.1978	0.2533

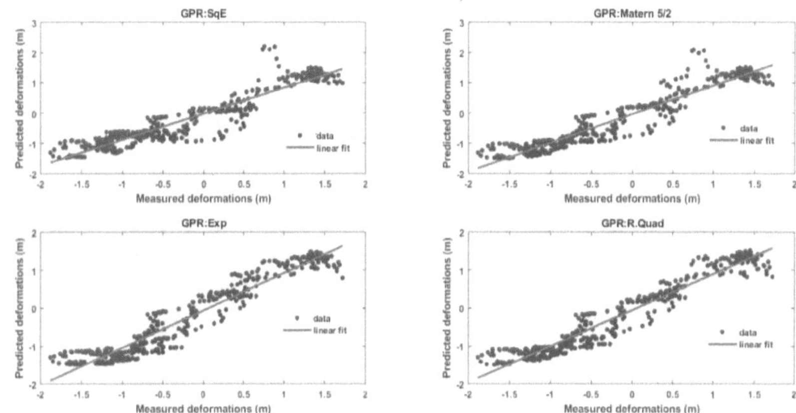

Figure 6.36. Predicted values versus measured values for radial deformation test data using GPR models

Table 6.24 presents the prediction performance of GPR models with different kernel functions for the tangential deformations. It can be observed that the GPR model based on Exp-based GPR model has the best performance in the training stage having the lowest MAE and RMSE values of 0.1164 and 0.1865 respectively. In the testing stage, Exp-based GPR model had the best performance with the lowest MAE and RMSE values of 0.3326 and 0.359 respectively. This implies that overall, Exp-based had the best performance among the kernels. Figure 6.35 shows the scatter plots of observed values and predicted values of the radial deformations using GPR models. Figure 6.35 indicates that values are distributed around the centre line.

Table 6.24. Prediction accuracy of tangential deformation based on GPR models using different kernels

Kernel	Training			Testing		
	R^2	MAE	RMSE	R^2	MAE	RMSE
SqE	0.96	0.1303	0.1961	0.83	0.3651	0.4336
Matern5/2	0.96	0.1269	0.1941	0.85	0.3298	0.3974
Exp	0.96	0.1164	0.1865	0.88	0.3326	0.359
R.Quad	0.96	0.1251	0.1929	0.88	0.3293	0.3647

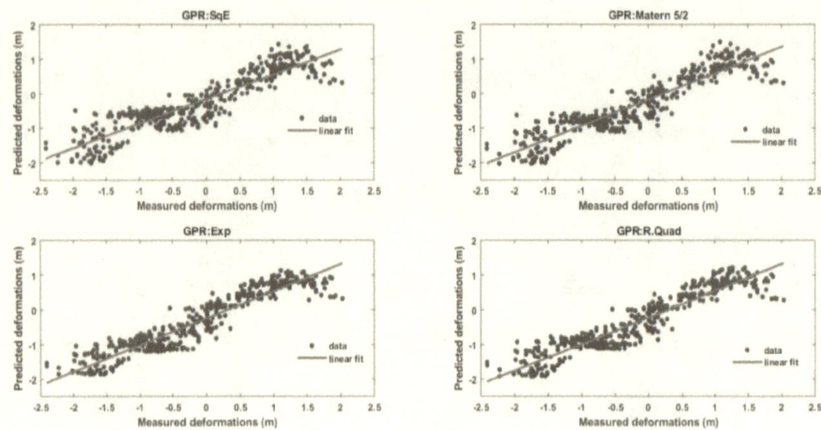

Figure 6.37. Predicted values versus measured values for vertical deformations test data using GPR models

Table 6.25 presents the prediction performance of GPR models with different kernel functions for the tangential deformations. It can be observed that the GPR model based on the Exp-based GPR model has the best performance in the training stage having the lowest MAE and RMSE values of 0.2657 and 0.3615 respectively. In the testing stage, Exp-based GPR model had the best performance with the lowest MAE and RMSE values of 0.6997 and 0.5656 respectively. This implies that overall, Exp-based had the best performance among the kernels. Figure 6.36 shows the scatter plots of observed values and predicted values of the radial deformations using GPR models. Figure 6.36 indicates that values are distributed around the centre line.

Table 6.25. Prediction accuracy of vertical deformations based on GPR models using different kernels

Kernel	Training			Testing		
	R^2	MAE	RMSE	R^2	MAE	RMSE
SqE	0.83	0.2729	0.3689	0.67	0.7174	0.6775
Matern5/2	0.83	0.2720	0.3681	0.72	0.6904	0.6291
Exp	0.84	0.2657	0.3615	0.77	0.6997	0.5656
R.Quad	0.83	0.2709	0.3677	0.75	0.6953	0.5859

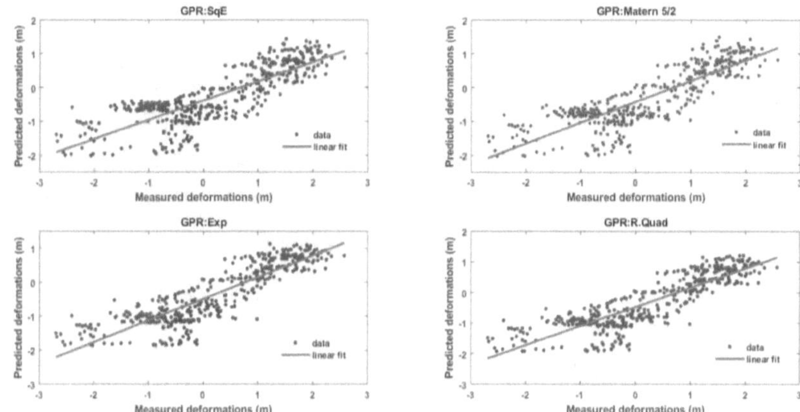

Figure 6.38. Predicted values versus measured values for vertical deformations test data using GPR models

One of the objectives of this research study was to understand the effect of water temperature on deformations. Similar procedures were followed in the prediction of natural frequencies using GPR models with water level and water temperatures as inputs in the regression analysis. Results obtained in the analysis are shown in Appendices A3 and A4. Results show that the prediction accuracy reduced by 12% for deformations. This showed that water temperatures do not have a significant effect on deformations.

The overall prediction performance of GPR models was tested on the entire set of deformations. Tables 6.26, 6.27 and 6.28 show the performance of these models in the prediction of radial, tangential and vertical deformations respectively. Results show that Exponential kernel function outperforms other kernels with the lowest RMSE values of 0.169, 0.279 in the prediction of radial and tangential deformations respectively. In the prediction of vertical deformation, R-Quad kernel function performed better with the lowest RMSE value of 0.379. Figures 6.37, 6.38 and 6.39 show a plot of observed and predicted values of the radial, tangential and vertical deformations respectively. These figures show that GPR predictions are close to the observed values.

Table 6.26. Prediction accuracy of GPR models using the whole data set for radial deformations

Kernel	Criteria		
	R^2	MAE	RMSE
SqE	0.97	0.132	0.186
Matern5/2	0.97	0.122	0.178
Exp	0.97	0.114	0.169
R.Quad	0.97	0.113	0.171

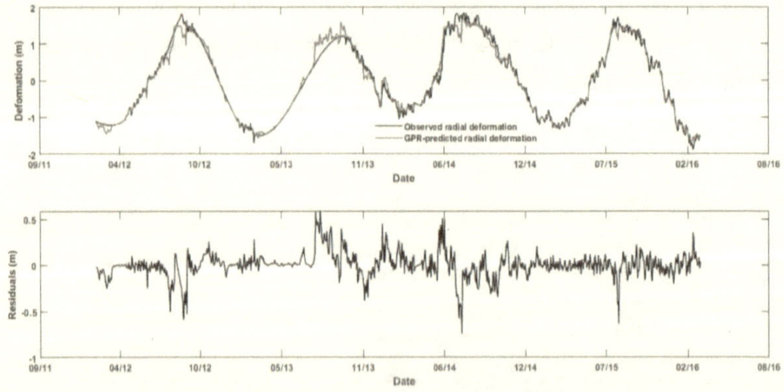

Figure 6.39. Observed and predicted radial deformations using GPR-Exp model with corresponding residuals

Table 6.27. Prediction accuracy of GPR models using whole data set for tangential deformations

Kernel	Criteria		
	R^2	MAE	RMSE
SqE	0.91	0.225	0.299
Matern5/2	0.91	0.219	0.293
Exp	0.92	0.205	0.279
R.Quad	0.92	0.207	0.282

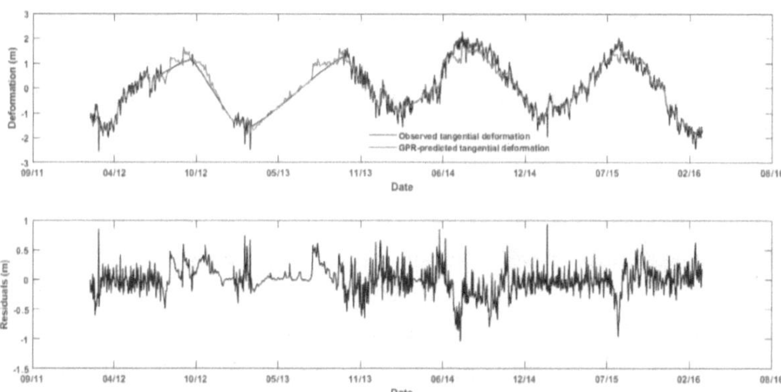

Figure 6.40. Observed and predicted tangential deformations using GPR-Exp model with corresponding residuals

Table 6.28. Prediction accuracy of GPR models using whole data set for vertical deformations

Kernel	Criteria		
	R^2	MAE	RMSE
SqE	0.84	0.2964	0.394
Matern5/2	0.85	0.2895	0.386
Exp	0.85	0.2854	0.381
R.Quad	0.86	0.2833	0.3787

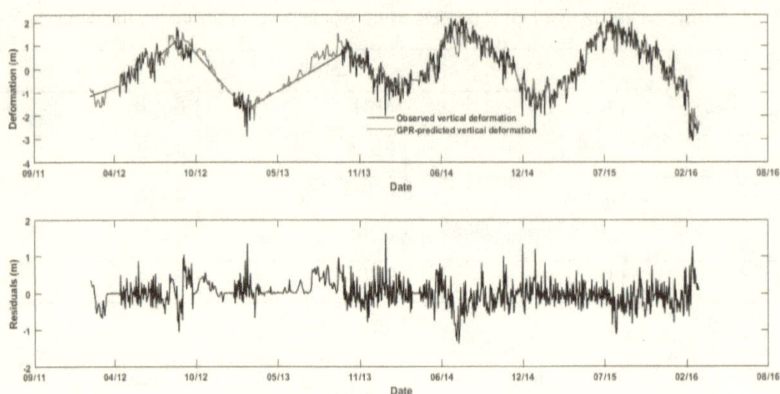

Figure 6.41. Observed and predicted vertical deformations using R-Quad model with corresponding residuals

6.3.4.2 Performance of other machine learning deformation models

Tables 6.29, 6.30 and 6.31 show the performance parameters namely R-squared, mean absolute error and root mean square error for selected machine learning models (SVR, BRT and ANN) for training and test data set for the radial, tangential and vertical deformations of Roode Elsberg Dam respectively.

Table 6.29. Performance of ML models on radial deformations

Model	Training			Testing		
	R^2	MAE	RMSE	R^2	MAE	RMSE
SVR	0.96	0.148	0.205	0.88	0.297	0.357
BRT	0.96	0.153	0.207	0.89	0.252	0.325
ANN	0.95	0.165	0.214	0.91	0.263	0.313

Table 6.30. Performance of ML models on tangential deformations

Model	Training			Testing		
	R^2	MAE	RMSE	R^2	MAE	RMSE
SVR	0.90	0.2219	0.300	0.77	0.419	0.493
BRT	0.91	0.2231	0.289	0.84	0.385	0.422
ANN	0.91	**0.2251**	**0.283**	0.82	0.403	0.457

Table 6.31. Performance of ML models on vertical deformations

Model	Training			Testing		
	R^2	MAE	RMSE	R^2	MAE	RMSE
SVR	0.82	0.274	0.378	0.67	0.644	0.70
BRT	0.83	0.287	0.368	0.75	0.658	0.611
ANN	0.83	0.275	0.325	0.74	0.572	0.487

The overall prediction performance of the selected ML models was tested on the entire set of natural frequencies. Tables 6.32, 6.33 and 6.34 show the performance of these models in the prediction of radial, tangential and vertical deformations respectively. Results indicate that ANN outperforms BRT and SVR because it has the lowest RMSE value of 0.180. Figures A6.1 to A6.6 show plots of observed and predicted values of the deformations using BRT, SVR and ANN models respectively. These figures show that SVR predictions are close to the observed values.

Table 6.32. Performance of ML models on radial deformations on full data

Kernel	Criteria		
	R^2	MAE	RMSE
BRT	0.96	0.156	0.206
SVR	0.96	0.140	0.192
ANN	0.96	0.131	0.180

Table 6.33. Performance of ML models on tangential deformations on full data

Kernel	Criteria		
	R^2	MAE	RMSE
BRT	0.91	0.229	0.298
SVR	0.91	0.219	0.297
ANN	0.91	0.224	0.285

Table 6.34. Performance of ML models on vertical deformations on full data

Kernel	Criteria		
	R^2	MAE	RMSE
BRT	0.80	0.345	0.444
SVR	0.82	0.308	0.418
ANN	0.81	0.333	0.402

6.3.4.3 Multivariate adaptive regression splines

MARS was used to predict deformations collected from the dam using a Matlab toolbox by Jacobsen (2016). Water level and air temperature were used as the inputs and deformations as the output.

Data was split into training and testing tests in the ratio of 80% and 20% respectively. In this analysis, the piecewise cubic model was used in the modelling of deformations as they provided a better approximation than the linear models. During the process of modelling using MARS, a set of BFs are selected by forward and backward iterations. Then the partitioned points are compared using the RSME in the iterations. Finally, the so-called modified generalised cross-validation (GCV) is used as a criterion to choose a final model. Also, MARS uses ANOVA to generate and compare models in terms of the importance of the inputs. Table 6.35 below shows the ANOVA decomposition selected using the GCV as a criterion for radial functions. The GCV column represents the significance of the corresponding ANOVA function by giving the GCV value for a model with all BFs corresponding to that particular ANOVA function to be calculated. Table 6.36 presents the performance of MARS models based on MAE and R^2 for training and testing data.

Table 6.35. ANOVA decomposition of the MARS model selected with GCV criterion

Function	Std	GCV	#Basis	#Params	Variable
1	0.1059	0.0064	2	5	1
2	15.2418	0.0082	3	7.5	2

Figure 6.40 shows a plot of how GCVs and MSE values change during the iterations. This figure shows that the lowest value of GCV occurs at 18 BFs. The same criterion was used for tangential and vertical deformations.

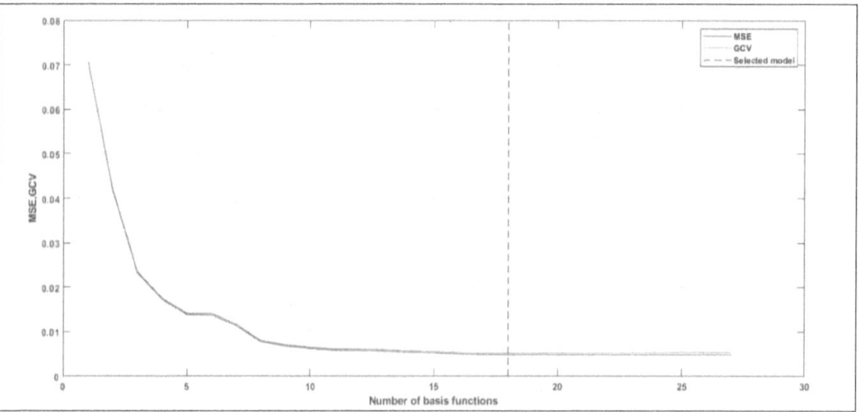

Figure 6.42. MARS model parameters

Six MARS models were developed to simulate deformations collected at Roode Elsberg Dam. The performance of these models based on correlation coefficient and mean absolute error for both training and testing data is summarised in Table 6.36. Results in this table show that MARS models were able to predict deformation to a good degree during the training phase besides 206E. However, the real performance of these models would only be known when subjected to test data.

MARS models performed better with radial deformations (206N) than the rest of the deformations. This is also shown with the MAE values as radial deformation (206N) has the lowest values.

Table 6.36. Performance of MARS developed models

Target/Criteria		Training			Test		
		R^2	MAE	RMSE	R^2	MAE	RMSE
Radial	dy1	0.92	0.208	0.2774	0.70	0.4395	0.655
	dy2	0.96	0.1436	0.1867	0.85	0.2988	0.4016
Tangential	dx1	0.92	0.2097	0.2671	0.83	0.3397	0.3939
	dx2	0.68	0.3883	0.4264	0.65	0.8434	0.7108
Vertical	dz1	0.84	0.2877	0.333	0.69	0.69	0.4939
	dz2	0.84	0.2704	0.3257	0.68	0.660	0.4817

The overall prediction performance of the selected MARS models was tested on the entire set of deformation. Table 6.37 shows the performance of these models in the prediction of radial, tangential and vertical deformations respectively. Figures 6.41, 6.42 and 6.43 show plots of observed and predicted values of the deformations using MARS models for radial, tangential and vertical deformations respectively. These figures show that MARS predictions are close to the observed values.

Table 6.37. Performance of the MARS model using the entire data set

Target/Criteria		Criteria		
		R^2	MAE	RMSE
Radial	dy1	0.91	0.214	0.290
	dy2	0.96	0.157	0.209
Tangential	dx1	0.90	0.204	0.309
	dx2	0.71	0.402	0.532
Vertical	dz1	0.78	0.372	0.464
	dz2	0.78	0.362	0.459

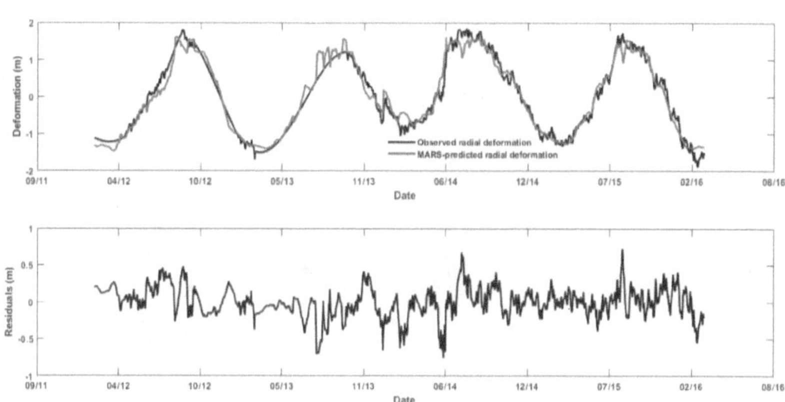

Figure 6.43. Observed and predicted radial deformations with residuals using the MARS model

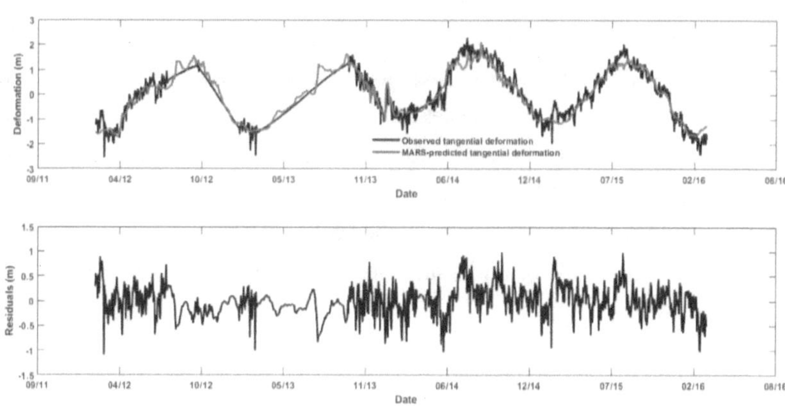

Figure 6.44. Observed and predicted tangential deformations with residuals using the MARS model

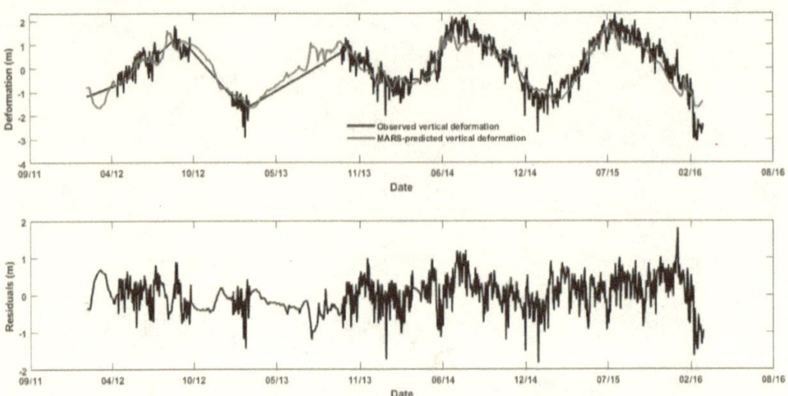

Figure 6.45. Observed and predicted vertical deformations with residuals using the MARS model

6.3.4.4 Multiple linear regression

In modelling of deformations from Roode Elsberg Dam, the known HST model described in Chapter 3 was used. In this study, only the effects due to hydrostatic and thermal loads were used in the model. Due to the lack of temperatures of the concrete dam wall, the thermal load was estimated using Equation 3.1. The time effect was not considered due to the shorter time used time series of the measured deformations. Deformations used in this analysis spanned from 01/01/2012 to 01/03/2016 resulting in 1492 observations. Using the least-squares method, the best MLR model was obtained using Equation 3.1. Table 6.38 shows the obtained regression coefficients.

Table 6.38. Performance of MLR developed models on deformations

Target/Criteria		Training			Test		
		R^2	MAE	RMSE	R^2	MAE	RMSE
Radial	dy1	0.83	0.0043	0.0054	0.91	0.0043	0.004
	dy2	0.94	0.0025	0.0032	0.93	0.0026	0.0027
Tangential	dx1	0.89	0.0009	0.0012	0.83	0.0015	0.0014
Vertical	dz1	0.79	0.0018	0.002	0.75	0.0046	0.0022
	dz2	0.77	0.0018	0.0019	0.72	0.0034	0.0022

Table 6.39. Regression coefficients and statistics from modelling

Parameter		dy1	dx1	dz1	dy2	dz2
Regression coefficients	β_0	0.0235	0.0032	0.0101	0.0228	0.0094
	β_1	-0.1487	-0.0353	-0.044	-0.1098	-0.0456
	β_2	0.4959	0.1179	0.1208	0.3328	0.1407
	β_3	-0.7669	-0.1758	-0.175	-0.4799	-0.1991
	β_4	0.4026	0.0903	0.0908	0.2403	0.0995
	β_5			0.0017	-0.0018	0.007
	β_6	-0.0087	-0.0032	-0.0036	-0.0117	-0.0036
	β_7		0.0012	0.002	0.0017	0.0009
	β_8			-0.0012	0.011	

Figures 6.44 shows the predicted and measured radial deformations using MLR models. As seen in these figures, MLR based models are close to the measured values.

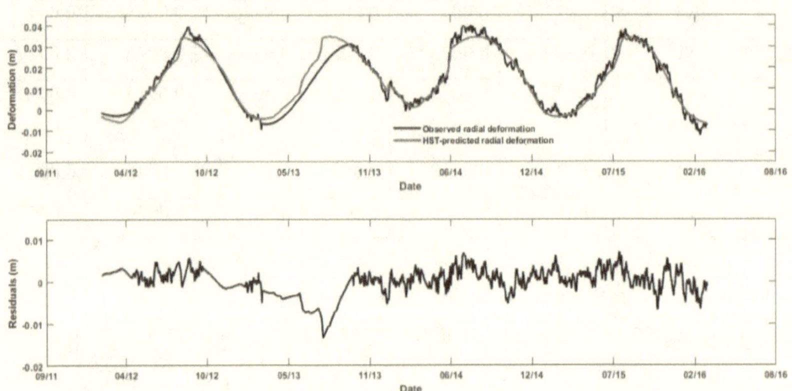

Figure 6.46. Observed and predicted radial deformation with residuals using the HST model

Figure 6.45 shows the effect of water level and temperature as predicted by the MLR model on the radial deformation. Figure 6.45 shows that temperature affects radial deformations more than water level. Observation of this figure indicates that water level contributes 20 mm of deformation while thermal load contributes 25mm.

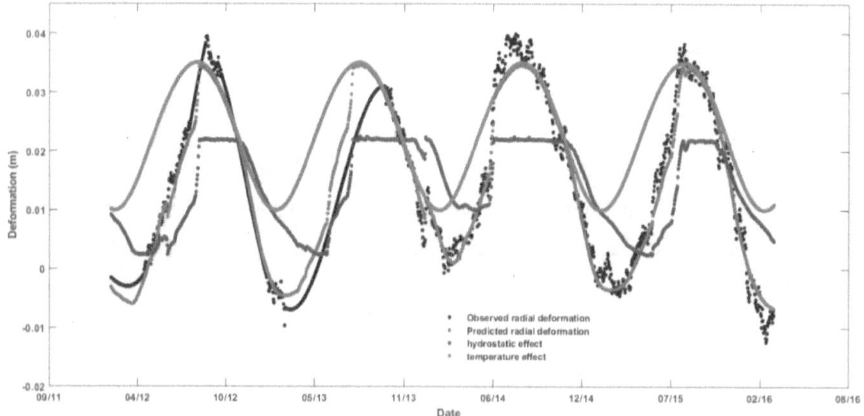

Figure 6.47. Radial deformation time series of measured and fitted values and its components by hydraulic and temperature model

In summary, machine learning models performed better than the MARS and MLR models in modelling both dam deformations and natural frequencies. Also, the inclusion of water temperature in the models improved the accuracy of deformation models which makes water temperatures another important factor to be considered while monitoring dams. Section 6.4 presents anomaly detection results obtained from the data.

6.4 Anomaly detection results

The identification of anomalies in dam monitoring data is of great significance in dam safety management. This section presents the results obtained from using the methodologies described above. Before robust techniques were applied to data, univariate methods were applied to data to ascertain some outliers by utilising residuals obtained from the constructed statistical models in Chapter 6. In this chapter, several statistical models were constructed to model deformations and natural frequencies for Roode Elsberg Dam. All models were able to predict the change in the dam structural responses (natural frequencies and deformations) within acceptable limits. Although all models performed well, Gaussian process models outperformed the other models. The first step is to compare model errors obtained from GP models with the mostly HST model. Outlier analysis based on robust statistics is then presented for both deformation and natural frequencies.

6.4.1 Natural frequency results

In dam surveillance, one of the objectives is to set thresholds such that if there is an event and these thresholds are exceeded then something must be done. Setting thresholds is normally done using residuals with assumptions that they are independent and normally distributed with mean zero and constant variance σ^2. The normal distribution assumptions can be checked using the normal probability plots of residuals (Walpole et al., 2012) obtained from the different models. Normal probability plots of residuals obtained from the GPR model and MLR are presented in Figures A5.1 and A5.2 (see Appendix A5).

Figures 6.46 and 6.47 show the univariate control charts based on GPR and MLR methods. Observations of these figures show that the majority of the observations lie in the established band with two and three observations outlying based on HST and BRT respectively. Examining outliers in Figures 6.49 and 6.50 show that outliers exist during the period when the water level was decreasing causing a scatter. Figure 6.50 further shows that there is an outlier at observation 590; this was due to a quick decrease in the water level leading to a slight increase in the natural frequency.

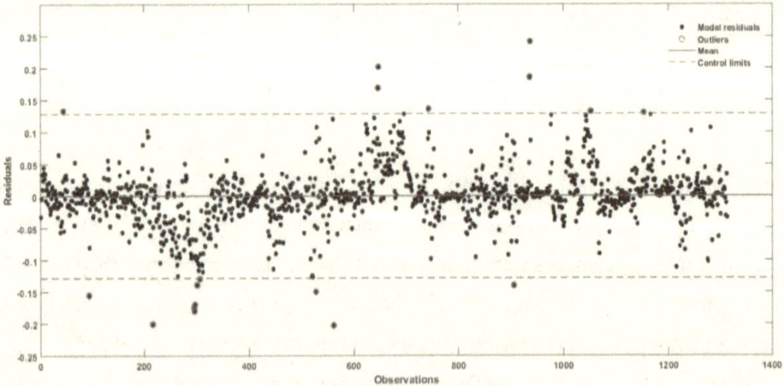

Figure 6.48. Univariate control chart for mode 1 based on the GPR model

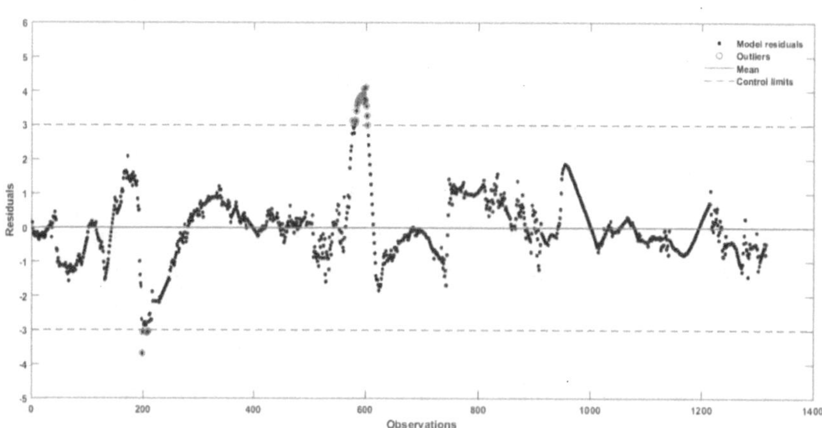

Figure 6.49. Univariate control chart for mode 1 based on the GPR model.

Figure 6.48 shows a regression outlier map, i.e. a plot of standardized LTS residuals and robust distance. Observation of this plot shows that there are three vertical outliers (small robust distance and large LTS-residual) with the rest being boundary values. The other observations on the right-hand side of the vertical line are good leverage points and follow the same pattern as the main group.

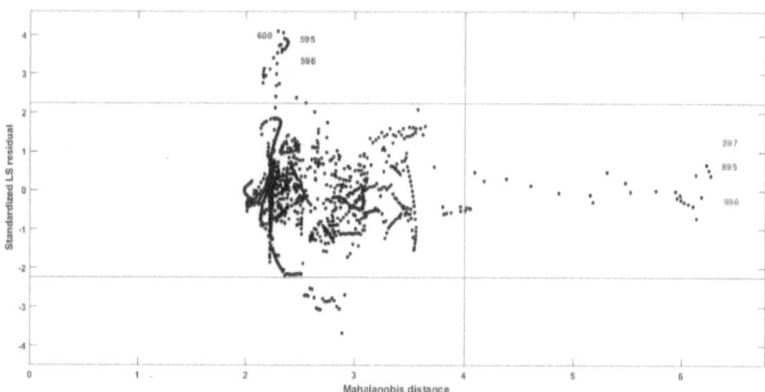

Figure 6.50. Regression diagnostic plot for mode 1

6.4.2 Deformation results

Univariate control charts for radial residual deformations resulting from the statistical models GPR, HST based on least trimmed squares are shown in Figures 6.49 and 6.50 respectively. Setting the threshold for residuals was based on the normality assumptions. Univariate control charts for model errors resulting from other models are presented in the Appendix. Observing Figure 6.49 shows that the largest band of model errors stay with the confidence bounds with outliers emerging at between 02/06/13 and 30/08/13. This period corresponded to the region of quickly rising water levels due to rains in Cape Town in the winter season. The dam wall is pushed downstream at a faster rate hence these appearing as outliers. On the other hand, the GPR models show outliers not only around the same dates as HST but also captures outliers between 22/04/14 to 21/8/14 which correspond to periods when there was a rise in the water level. This was not captured by the HST period as the outliers obtained correspond to the periods of temperature was low. Examining outliers in Figure 6.52 shows that outliers exist due to the interpolated data hence false alarms. These figures show that apart from a few outliers, the residuals are within the confidence limits. This shows that the GPR model can be utilised as an indicator of the dam's structural condition. Generally, these figures show that the dam behaved normally during this period.

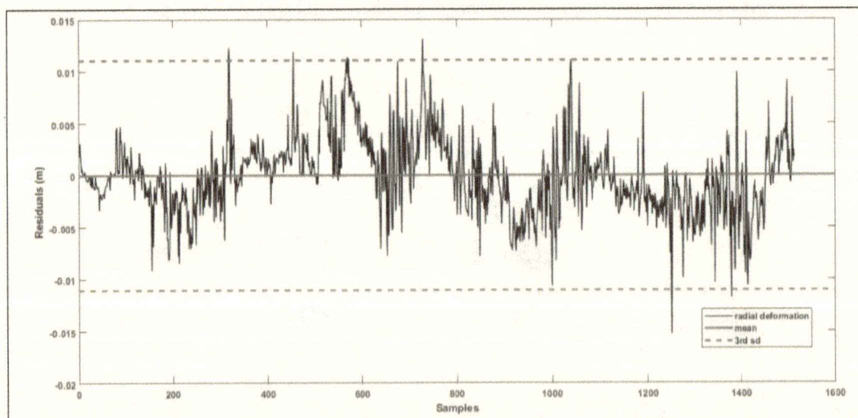

Figure 6.51. Univariate control chart for radial deformations based on GPR

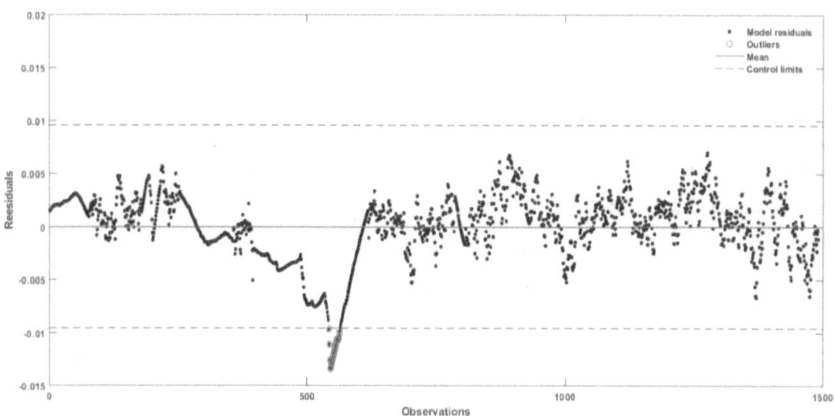

Figure 6.52. Univariate control chart for radial deformations based on HST

Figure 6.51 shows a regression outlier map, i.e. a plot of standardized LTS residuals and robust distance. Observation of this plot shows that there are three vertical outliers (small robust distance and large LTS-residual) with the rest being boundary values. The other observations on the right-hand side of the vertical line are good leverage points and follow the same pattern as the main group.

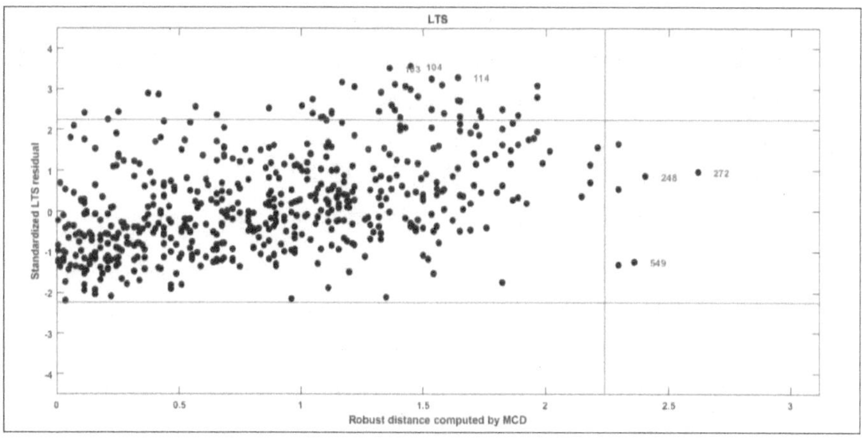

Figure 6.53. Regression diagnostic plot for radial deformation

6.5 Chapter summary

This chapter presents the continuous monitoring (ambient vibration and deformation) of Roode Elsberg Dam. Plots of environmental loads (water level, air and water temperature) and the dam responses (deformations and natural frequencies) were made. The following key observations made from the continuous monitoring data used are;

- The maximum vertical deformations of Roode Elsberg Dam are about 2 times smaller than the maximum radial deformations.
- Tangential deformations of Roode Elsberg Dam move in opposite directions due to different ground conditions on either side of the spillway.
- The seasonal deformations of the dam are strongly dependent on temperature variation. While the seasonal variation of deformations is largely driven by temperature variation, the reservoir level does influence the deformations of the dam especially as the dam fills rapidly.
- Radial and tangential deformations are more influenced by changes in reservoir water level than vertical deformations.
- Water level changes influence the tangential deformations but not as much as temperature
- The natural frequencies of the dam are strongly dependent on the reservoir level. The natural frequencies of the dam decrease with the water level. The dam's natural frequencies are weakly influenced by temperature variation. However, during periods of constant water level, the effect of temperature was seen, i.e., an increase in temperature led to a slight increase in the natural frequency.
- At high temperatures and low water levels, the scatter in the natural frequencies is large.

The main objective of this study was to understand the relationship between environmental loads and the natural frequencies and deformations. A machine learning-based predictive model, namely Gaussian Process regression, which supports vector regression, boosted regression trees and artificial neural networks was developed to understand the interactions. Furthermore, the performance of the machine learning-based model was compared with other models namely; multiple linear regression and multivariate adaptive regression splines to predict both natural frequencies and deformations of Roode Elsberg Dam using measurements of water level, temperature (air and water). The following conclusions were made.

- Multiple linear regression has been most used in the prediction of deformation of dams. In this chapter, MLR based on least squares and MLR based on least trimmed squares were tested in the prediction of deformations and natural frequencies and it was possible to distinguish hydrostatic and thermal effects.
- Machine learning algorithms are the most appropriate methodologies in the prediction of both deformations and natural frequencies of the dam. Among the machine learning algorithms, Gaussian Process Regression produced better results with the lowest value of MAE with deformations and boosted regression trees being better in the prediction of natural frequencies.
- Multivariate Adaptive Regression Splines were used in the prediction of both deformation and natural frequencies using temperature and water level as model inputs. Results showed that MARS is a promising algorithm that can be used in dam monitoring.
- Results showed that it was easier to predict radial deformations than tangential and vertical deformations. This is attributed to the higher signal-noise ratio in the vertical and tangential deformations.
- Results from control charts obtained from the commonly used HST and the proposed GPR were compared. Observations have shown that GPR detects more outliers than HST for deformations and BRT for natural frequencies. Least trimmed squares method uses robust statistics for anomaly detection. The robust methods were able to detect outlying observations that the control charts would not detect.

7 Summary, conclusions and recommendations

7.1 Introduction

This work focused on understanding the influence of environment and operational loads (reservoir water level and temperature) on the dam structural responses namely deformations and natural frequencies through the development of statistical models. The main case study of this book is the monitoring campaign of Roode Elsberg Dam located in Worcester, Western Cape, South Africa. The main objectives of this study were: -

(i) Establishing the relationships between loading (water level and temperature) and the dynamic properties (natural frequencies)

(ii) Predicting natural frequencies and deformations from monitoring data using data-driven models

(iii)Using the results obtained in objectives (ii) for detecting anomalies.

To achieve the abovementioned objectives, a comprehensive literature review was made focusing on what methodologies were used in deformation and ambient vibration monitoring. Gaps in the literature were identified from which statistical models mainly focusing on Gaussian process regression models were developed to establish relationships between dam responses and environmental variables collected from the monitoring systems installed on Roode Elsberg Dam. Therefore, this chapter summarizes the key issues presented in this work, conclusions and recommendations.

7.2 Summary

The purpose of any dam monitoring scheme is to ensure the safety of dam structures. Chapter 2 introduced the concept of dam safety where much emphasis was put on risk-based approaches in ensuring dam safety. Chapter 3 reviews the case studies where monitoring schemes have been done on concrete dams. Methodologies used in the analysis of the collected data are also described citing their advantages and disadvantages. This review was important in the selection of the methodologies to be used in the analysis of data collected from the monitoring campaign.

Chapter 4 builds on the knowledge built from Chapter 3 in the choice of methodologies that were used for the analysis of dam monitoring data. In this chapter, Gaussian Process regression and a robust based multiple linear model based least trimmed square are described. In GPR, the major focus was on the choice of the kernel function in the process of developing models based on GPR. The purpose of developing statistical models in dam engineering is to detect anomalies, methodologies based on control charts and least trimmed squares, which are described in Chapter 4.

Chapter 5 described both continuous monitoring systems that are installed on the dam. These include (i) the Global Navigation satellite system (GNSS) with two sensors placed on the dam crest and two sensors near the dam site. These sensors measure deformations of the dam in three directions (x, y, z) which are processed in a computer located at the dam site at a daily rate. The dynamic monitoring system consists of three forced accelerometers placed on the dam crest and a data acquisition system placed in the upper gallery. The dynamic monitoring system is used to monitor ambient vibrations due to water waves, wind and ground motion. Also, there is a weather station located on the dam site to measure air temperature thermocouples that are inserted in the dam wall to monitor the water temperature. The reservoir water level is measured by an automated system located on the dam site. Results obtained from the monitoring schemes are discussed in Chapter 6.

7.3 Conclusions

This book investigated temperature and water level effects on the dam responses (deformations and natural frequencies) of Roode Elsberg Dam through the development of regression models. Dam deformations were measured using the GNNS and natural frequencies were estimated from the measured accelerations using an operational modal analysis technique called data-driven stochastic subspace identification method. The following conclusions were made from the study.

From the literature review, it can be seen that monitoring of dams has been going on for more than 40 years. Much focus has been on visual inspections and periodic tests where dam deformations are measured through geodetic surveys. In terms of continuous monitoring, deformation monitoring systems are mostly installed systems on dams and a lot of work has been carried out in this regard. From the mid-1980s, periodic dynamic tests in the form of forced and ambient vibration tests have been conducted on dams such that dynamic calibration of numerical models. In this regard, periodic ambient vibration tests were carried out on Roode Elsberg Dam between 2008 and 2013. Results show that three dominant natural frequencies in the range of 3Hz and 6Hz were estimated. These results were used as a baseline for the installation of a continuous ambient-based monitoring system on the dam.

Following the geodetic and ambient vibration surveys, two continuous monitoring systems (dynamic and deformation) were installed on the dam. Continuous monitoring tracks the variation of both natural frequencies and deformations over time and also helps in understanding the effects of environmental variables (temperature and water level) on the dam responses. Monitoring results indicated that there are three modes in the range of 3Hz-6Hz which were continuously monitored. Observations showed that the water level is the driving factor in the variation of dam natural frequencies. There exists an inverse relationship between water level and natural frequencies, i.e., when the water level rises, natural frequencies reduce since there is an added mass on the dam and they increase with a reduction in water level. The effect of air temperature is at a period when the dam is spilling and the water level is constant. During this period, there was a slight increase in the natural frequencies attributed to an increase in temperature. Water temperatures did not show any effect on natural frequencies of the dam.

Results from the deformations showed that the maximum radial, tangential and vertical deformations were 40mm, 15mm. and 10mm respectively. Maximum radial deformations were twice the maximum vertical deformations. Observations showed that temperature is the major driving factor in the change of deformations followed by water level. When temperatures are high, dams tend to contract causing the dam wall to move upstream and inwards while when the temperature is low, dams expand pushing the dam wall downstream. Similarly, when the water level rises, there is an added mass on the dam wall, pushing the wall downstream and when the water level goes down, the wall moves back upstream. There was an interesting observation during the periods of the rapid rise of water level which resulted in a dam wall moving downwards and then back. Furthermore, a change of about 20 degrees in temperature produced a deformation of about 25mm.

While observing the effect of water level and temperature effects on natural frequencies and deformations, it was imperative to develop relationships between these variables using statistical models. Due to the nonlinear relationships between the input and output variables, it was important to develop statistical models that would yield good results. In this regard, Gaussian process regression models were proposed to predict the variation of both natural frequencies and deformation using water level and temperature as predictors. GPR based models were chosen to establish a relationship between environmental variables and specifically natural frequencies as, to the best of my knowledge, they have not been used in this regard. Furthermore, GPR models have advantages of dealing with nonlinearity and production of the predictive distribution. The challenge in GPR modelling is the choice of a covariance function and the corresponding hyperparameters since every covariance function is unique. Results showed that for natural frequency modelling, the exponential based covariance function performed best with R^2, RMSE and MAE values 0.99, 0.081, 0.054 respectively. For deformation modelling the exponential based covariance function performed best with R^2, RMSE and MAE values of 0.97, 0.169 and 0.114 respectively for radial deformations. The performance of the GPR models was compared with other models namely MARS, BRT, ANN, SVR, HST using performance indicators including R^2, RMSE and MAE. Generally, results showed that predicted results by GPR models agreed with the observed data and GPR deals with nonlinear behaviour of the dam and improved prediction accuracy.

Although dams can behave normally for many years, abnormal behaviour can occur even after many years of the normal behaviour of dams, and there can still be abnormal occurrences at any time. This makes it important to detect any abnormal behaviour in the collected data such that decisions concerning dam safety can be made. To this effect, control charts were constructed from model residuals from the best performing model. Results indicated that the dam behaved normally during the period of analysis. Distance-distance plots were also plotted and the same results were obtained as in control charts. The observed abnormalities were all false alarms.

7.4 Recommendations for future work

The presented work in this book included the development of statistical models to predict dam responses using data collected from two monitoring schemes at Roode Elsberg Dam. However, some pending issues are surrounding the monitoring of dams that still require more to be understood about the structural behaviour of dams. These include:

- Thermometers need to be installed on the dam wall (upstream and downstream) such that real temperature values can be used in the analysis of dam responses especially natural frequencies.
- Further research should be done concentrating on developing methodologies based on failure modes of dams, i.e., instruments and data analysis that can detect several failure modes of dams.
- In dynamic monitoring of dams, there is a need to install accelerometers on all the blocks of Roode Elsberg Dam such that full mode shapes can be obtained, and damping ratios can be monitored too for damage detection.
- Observations showed that natural frequencies are affected strongly by the change in water level at the dam. These can be used to approximate the mass of hydrostatic load on the dam wall.
- There is a need to develop a real-time online structural health monitoring system for Roode Elsberg Dam to detect any structural changes on the dam timely and identify any extreme events such as floods and long periods of drought.

APPENDICES

APPENDIX A

A.1 Prediction accuracy of GPR models on natural frequencies using water temperatures

Table A1. 1. Prediction accuracy of 1st natural frequency based on GPR models using different kernels with water temperatures

Kernel/Criteria	Training set			Testing set		
	R^2	MAE	RMSE	R^2	MAE	RMSE
SqE	0.94	0.1367	0.2525	0.74	0.3957	0.4833
Matern 5/2	0.94	0.1298	0.2248	0.80	0.3602	0.4335
Exp	0.95	0.1154	0.2343	0.87	0.3058	0.3589
R Quad	0.94	0.1207	0.2409	0.87	0.3021	0.3595

Table A1. 2. Prediction accuracy of 2nd natural frequency based on GPR models using different kernels with water temperatures

Kernel/Criteria	Training set			Testing set		
	R^2	MAE	RMSE	R^2	MAE	RMSE
SqE	0.92	0.1576	0.2780	0.71	0.3903	0.5265
Matern 5/2	0.93	0.1489	0.2715	0.76	0.3701	0.4859
Exp	0.94	0.1327	0.2529	0.83	0.3475	0.4023
R Quad	0.93	0.1432	0.2663	0.84	0.3504	0.3720

Table A1.3. Prediction accuracy of 3rd natural frequency based on GPR models using different kernels with water temperatures

Kernel/Criteria	Training set			Testing set		
	R^2	MAE	RMSE	R^2	MAE	RMSE
SqE	0.93	0.1675	0.2714	0.78	0.3418	0.4285
Matern 5/2	0.94	0.1589	0.2643	0.82	0.3142	0.3980
Exp	0.94	0.1433	0.2476	0.86	0.2902	0.3556
R Quad	0.94	0.1466	0.2533	0.86	0.2982	0.3409

A.2 Prediction accuracy of ML models on natural frequencies using water temperatures

Table A2 1. Performance of ML models on 1st natural frequency with water temperatures

Model	Training set			Testing set		
	R^2	MAE	RMSE	R^2	MAE	RMSE
BRT	0.92	0.1812	0.2869	0.89	0.2701	0.3202
SVR	0.92	0.1568	0.2920	0.80	0.3531	0.4156
ANN	0.93	0.1796	0.2597	0.87	0.3070	0.3609

Table A2 2. Performance of ML models on 2nd natural frequency with water temperatures

Model	Training set			Testing set		
	R^2	MAE	RMSE	R^2	MAE	RMSE
BRT	0.90	0.2125	0.3251	0.85	0..3342	0.3455
SVR	0.90	0.1715	0.3249	0.83	0.3491	0.3892
ANN	0.88	0.2177	0.3206	0.85	0.3608	0.3826

Table A2 3. Performance of ML models on 3rd natural frequency with water temperatures

Model	Training set			Testing set		
	R^2	MAE	RMSE	R^2	MAE	RMSE
BRT	0.90	0.2327	0.3327	0.87	0.2887	0.3318
SVR	0.91	0.1734	0.3052	0.83	0.3187	0.3850
ANN	0.89	0.2165	0.3153	0.87	0.3024	0.3574

A.3 Prediction accuracy of GPR models on deformations using water temperatures

Table A3 1. Prediction accuracy of radial deformations at P203 based on GPR models using different kernels with water temperatures

Kernel/Criteria	Training set			Testing set		
	R^2	MAE	RMSE	R^2	MAE	RMSE
SqE	0.95	0.1329	0.2281	0.78	0.4149	0.5518
Matern 5/2	0.96	0.1154	0.2001	0.88	0.2683	0.3588
Exp	0.98	0.0829	0.1459	0.90	0.3395	0.3590
R Quad	0.98	0.0925	0.1573	0.93	0.2890	0.2805

Table A3 2. Prediction accuracy of radial deformations at P206 based on GPR models using different kernels with water temperatures

Kernel/Criteria	Training set			Testing set		
	R^2	MAE	RMSE	R^2	MAE	RMSE
SqE	0.97	0.1099	0.1575	0.86	0.2951	0.38944
Matern 5/2	0.98	0.0932	0.1471	0.89	0.2516	0.3431
Exp	0.99	0.0687	0.1047	0.91	0.2281	0.3115
R Quad	0.99	0.0729	0.1094	0.92	0.2080	0.2740

Table A3 3. Prediction accuracy of tangential deformation (P203) based on GPR models using different kernels with water temperatures

Kernel/Criteria	Training set			Testing set		
	R^2	MAE	RMSE	R^2	MAE	RMSE
SqE	0.94	0.1686	0.2383	0.71	0.4697	0.5615
Matern 5/2	0.94	0.1632	0.2292	0.78	0.4097	0.4754
Exp	0.95	0.1465	0.2109	0.85	0.3322	0.3990
R Quad	0.95	0.1504	0.2172	0.86	0.3211	0.3750

Table A3 4. Prediction accuracy of tangential deformation (P206) based on GPR models using different kernels with water temperatures

Kernel/Criteria	Training set			Testing set		
	R^2	MAE	RMSE	R^2	MAE	RMSE
SqE	0.82	0.2796	0.3923	0.33	0.8986	0.9849
Matern 5/2	0.83	0.2681	0.3815	0.39	0.7906	0.7645
Exp	0.85	0.2453	0.3567	0.52	0.6788	0.6126
R Quad	0.84	0.2498	0.3619	0.66	0.6347	0.4132

Table A3.5. Prediction accuracy of vertical deformation (P203) based on GPR models using different kernels with water temperatures

Kernel/Criteria	Training set			Testing set		
	R^2	MAE	RMSE	R^2	MAE	RMSE
SqE	0.88	0.2243	0.3119	0.63	0.6455	0.5747
Matern 5/2	0.89	0.2117	0.296	0.68	0.6531	0.5163
Exp	0.91	0.1927	0.2744	0.71	0.6758	0.5208
R Quad	0.91	0.1944	0.2749	0.74		

Table A3 4. Prediction accuracy of vertical deformation (P206) based on GPR models using different kernels with water temperatures

Kernel/Criteria	Training set			Testing set		
	R^2	MAE	RMSE	R^2	MAE	RMSE
SqE	0.87	0.2307	0.3193	0.57	0.6036	0.5658
Matern 5/2	0.88	0.2176	0.3028	0061	0.5934	0.5302
Exp	0.91	0.1969	0.2739	0.62	0.6169	0.6219
R Quad	0.90	0.2003	0.2749	0.67	0.5974	0.4963

A.4 Prediction accuracy of ML models on deformations using water temperatures

Table A4. 1. Prediction accuracy of radial deformation (P203) based on ML models with water temperatures

Model	Training set			Testing set		
	R^2	MAE	RMSE	R^2	MAE	RMSE
BRT	0.94	0.1821	0.2499	0.89	0.3548	0.3271
SVR	0.94	0.1496	0.2399	0.87	0.2554	0.3515
ANN	0.94	0.1557	0.2339	0.79	0.4658	0.5769

Table A4. 2. Prediction accuracy of radial deformation (P206) based on ML models with water temperatures

Model	Training set			Testing set		
	R^2	MAE	RMSE	R^2	MAE	RMSE
BRT	0.97	0.1353	0.1824	0.87	0.2983	0.3630
SVR	0.97	0.1222	0.1745	0.89	0.2632	0.3112
ANN	0.94	0.1729	0.2273	0.92	0.2275	0.2780

Table A4.3. Prediction accuracy of tangential deformation (P203) based on ML models with water temperatures

Model	Training set			Testing set		
	R^2	MAE	RMSE	R^2	MAE	RMSE
BRT	0.92	0.2115	0.2700	0.88	0.3209	0.2996
SVR	0.94	0.1176	0.2471	0.77	0.4091	0.4342
ANN	0.93	0.1861	0.2423	0.83	0.3539	0.4687

Table A4. 5. Prediction accuracy of tangential deformation (P206) based on ML models with water temperatures

Model	Training set			Testing set		
	R^2	MAE	RMSE	R^2	MAE	RMSE
BRT	0.74	0.3492	0.4653	0.67	0.6109	0.4118
SVR	0.79	0.2977	0.4218	0.35	0.8567	0.7177
ANN	0.78	0.3229	0.3776	0.34	0.9744	0.9434

Table A4. 6. Prediction accuracy of vertical deformation (P203) based on ML with water temperatures

Model	Training set			Testing set		
	R^2	MAE	RMSE	R^2	MAE	RMSE
BRT	0.85	0.2681	0.3471	0.78	0.7384	0.3625
SVR	0.87	0.2268	0.3207	0.64	0.6731	0.5016
ANN	0.87	0.2427	0.2984	0.66	0.6238	0.6289

Table A4.7. Prediction accuracy of vertical deformation (P206) based on ML with water temperatures

Model	Training set			Testing set		
	R^2	MAE	RMSE	R^2	MAE	RMSE
BRT	0.85	0.2678	0.3428	0.72	0.6287	0.4142
SVR	0.86	0.2349	0.3283	0.57	0.6128	0.5128
ANN	0.85	0.2526	0.3108	0.61	0.6402	0.6200

A5: Natural frequency results

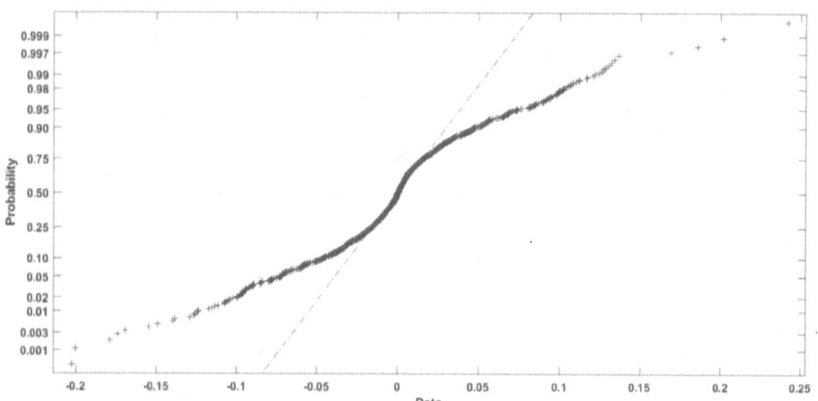

Figure A5. 1. Normal probability plot for mode 1 residuals based on GPR model

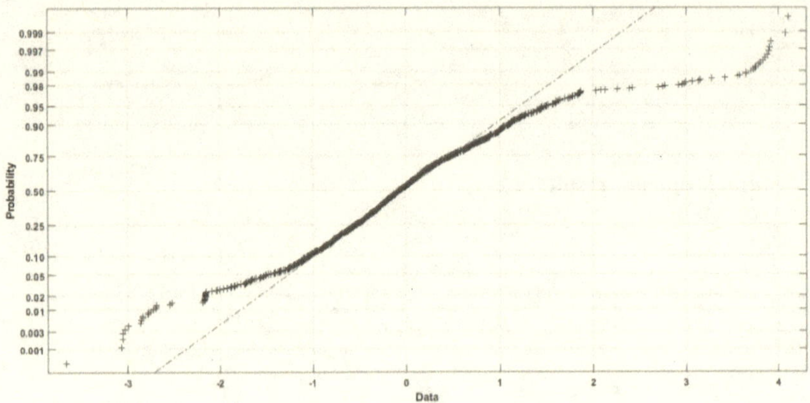

Figure A5. 2. Normal probability plot for mode 1 residuals based on GPR model

A6: Dam deformation results

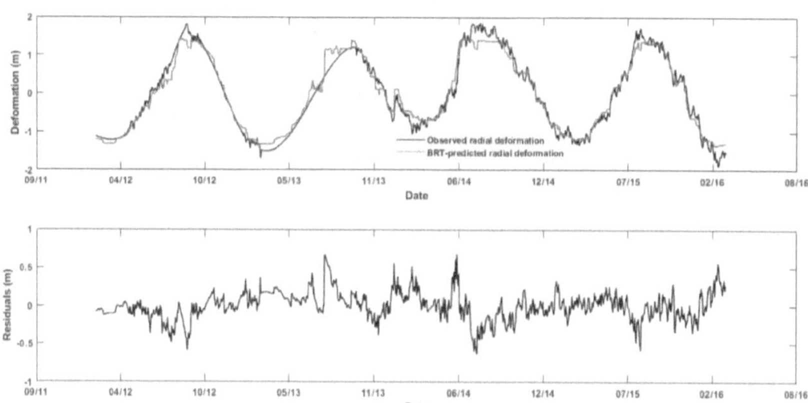

Figure A6.1 Measured and predicted radial deformation using BRT models and residuals

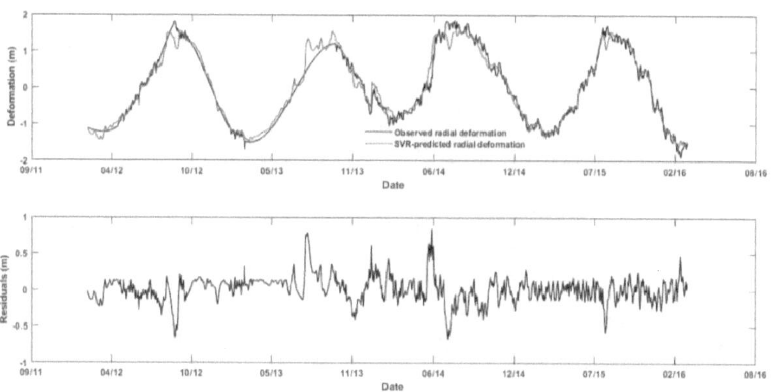

Figure A6. 2. Measured and predicted radial deformation using SVR models and residuals

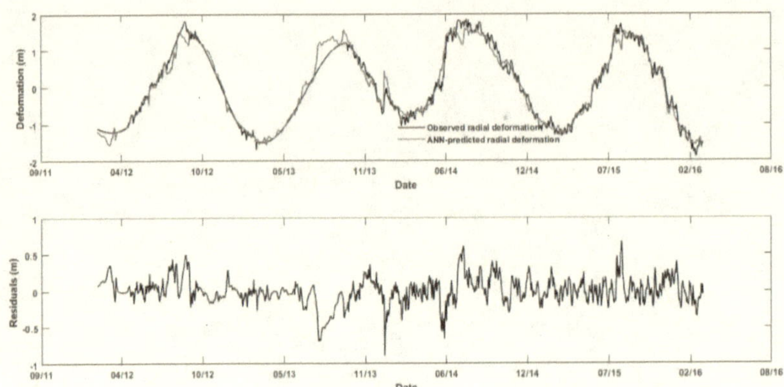

Figure A6. 3. Observed and predicted radial deformations using ANN models and residuals

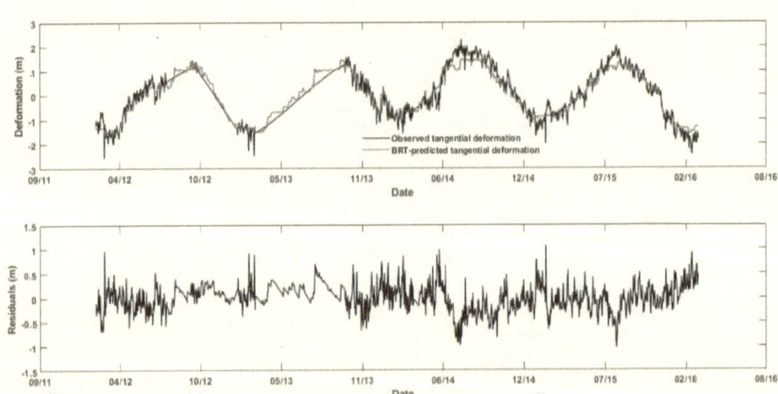

Figure A6. 4. Measured and predicted tangential deformations using BRT models and residuals

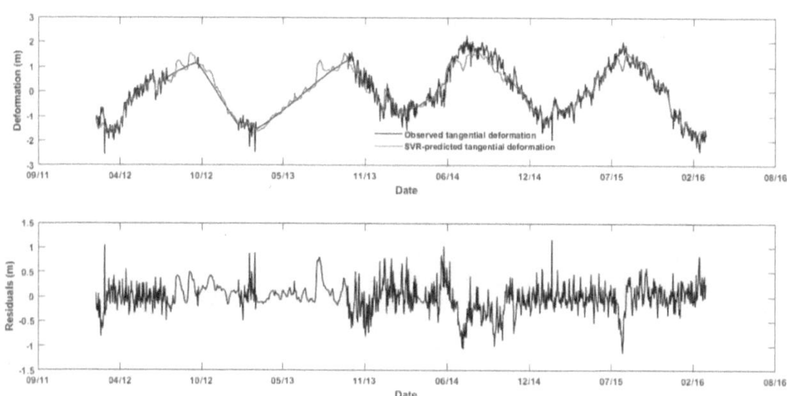

Figure A6. 5. Measured and predicted tangential deformations using SVR models and residuals

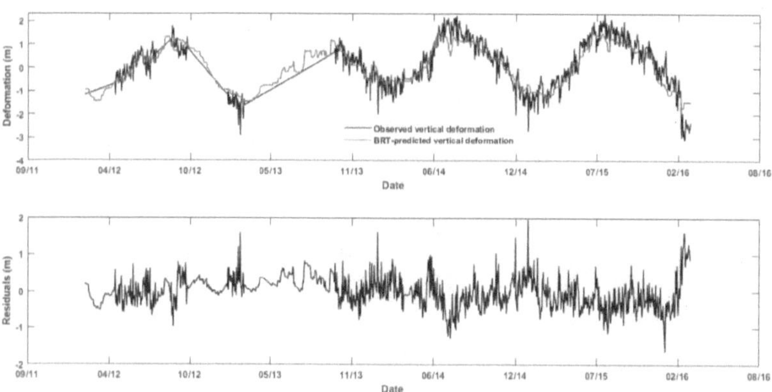

Figure A6. 6. Measured and predicted vertical deformations using BRT models and residuals

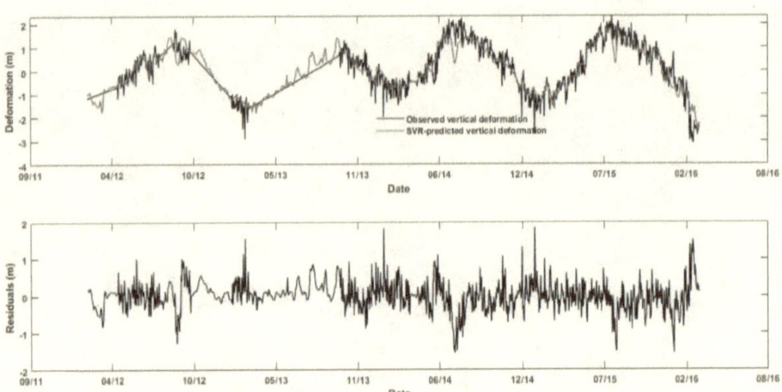

Figure A6. 7. Measured and predicted vertical deformations using SVR models and residuals

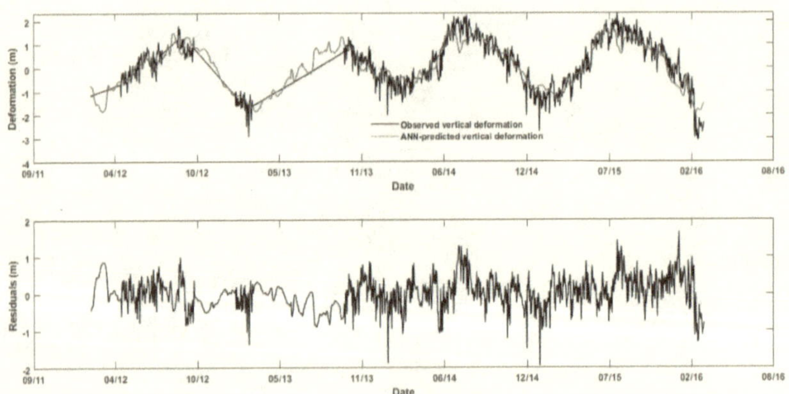

Figure A6. 7. Observed and predicted vertical deformations using ANN models and residuals

APPENDIX B

Author publications to date

Journal papers

- Bukenya, P, Moyo P and Beushasuen H (2014). Health monitoring of concrete arch dams: A literature review. J Civil Struct Health Monit. doi 10. 1007/s13349-014- 0079-2

Reviewed conference papers

- Bukenya, P, Moyo P, Oosthuizen, C (2012). Modal parameter estimation from ambient vibration measurements of a dam using stochastic subspace identification methods. In Proceedings of Concrete, Repair, Rehabilitation and Retrofitting, Taylor and Francis
- Bukenya P, Moyo P and Oosthuizen C (2014). Towards long term dynamic monitoring of Roode Elsberg dam. SANCOLD 2014, Johannesburg, 2014
- Bukenya P, Moyo P and Oosthuizen C (2016). Long term integrity monitoring of a concrete arch dam using continuous dynamic measurements and a multiple linear model. Proceedings of the International Symposium on Appropriate technology to ensure proper Development, Operation and Maintenance of Dams in Developing countries. Johannesburg, South Africa, 18 May 2016
- Bukenya P, Moyo P (2017). Monitoring the structural behaviour of concrete dams. The case of Roode Elsberg Dam, South Africa. In Proceedings of SANCOLD 2017, Pretoria, South Africa

Conference papers

- Bukenya, P, Moyo, P, Beushausen, H and Oosthuizen, C (2012). Comparative study of operational modal analysis techniques using ambient vibration measurements of a concrete dam. Proceedings of the 25th International Conference on Noise and Vibration Engineering, Leuven. Belgium: 17-19 September 201

- Bukenya P, Moyo P and Oosthuizen C (2013). Experimental modal identification of a South African Concrete arch dam. Proceedings of the 5th International Conference on Experimental Vibration Analysis for Civil Engineering Structures (EVACES) 29-30 Oct 2013 Ouro Preto, Brazil

- Moyo P and Bukenya P (2017). Experiences with continuous monitoring of deformation and modal properties of an arch dam. Proceedings of the 8th International Conference on Structural Health Monitoring of Intelligent Infrastructure Brisbane, Australia | 5-8 December 2017

www.ingramcontent.com/pod-product-compliance
Lightning Source LLC
LaVergne TN
LVHW091413190726
843491LV00006B/1407

* 9 7 8 3 3 8 4 2 2 4 2 7 9 *